Sebastian Herbstreuth completed his PhD in Politics and International Studies at the University of Cambridge and Georgetown University. He now works as a non-profit sector professional and independent author in Germany.

"America's ambivalent relationship with oil – its fears of foreign dependence and its wholehearted embrace of the hydrocarbon economy – has been brilliantly described in this study of the role that oil has played in the country's self-image and its foreign policy. Sebastian Herbstreuth has not limited his story to the technicalities of the subject but has, instead, sought to elucidate the inner meaning of America's embrace of oil and the international oil industry which it dominates. His book provides a deeply satisfying and insightful analysis of a political romance that profoundly influences global politics, an influence that seems likely to endure despite the lowering reality of climate change."

George Joffé is Senior Fellow of the Centre of International Studies at the Department of Politics and International Studies (POLIS) at the University of Cambridge

OIL AND AMERICAN IDENTITY

A Culture of Dependency and US Foreign Policy

SEBASTIAN HERBSTREUTH

I.B.TAURIS
LONDON · NEW YORK

Published in 2016 by
I.B.Tauris & Co. Ltd
London • New York
www.ibtauris.com

An earlier version of Chapter 4 is published in Sebastian Herbstreuth, "Constructing dependency: the United States and the problem of foreign oil", *Millennium Journal of International Studies* 43/1 (2014), pp. 24–42.

Library of Modern American History 8

ISBN: 978 1 78453 149 2
eISBN: 978 0 85773 838 7
ePDF: 978 1 78673 991 9

A full CIP record for this book is available from the British Library
A full CIP record is available from the Library of Congress

Library of Congress Catalog Card Number: available

Typeset in Garamond Three by OKS Prepress Services, Chennai, India

Für Emma

CONTENTS

LIST OF ILLUSTRATIONS

ACKNOWLEDGEMENTS

It is only appropriate to begin a book about dependency by acknowledging one's own reliance on the generous support and expertise offered by others. This book would have been impossible without the support of George Joffé at Cambridge University, who first encouraged me to explore the reservoir of crude ideas about oil that I had stumbled upon and then tirelessly helped me refine and market the products. I remain profoundly grateful to him.

In preparing this book, I was fortunate to be able to speak to a number of experienced experts in the fields of energy, American history and foreign policy. In particular, I would like to thank Tarak Barkawi, Matthew Darby, Stefan Halper, Gal Luft, Melani McAlister, David Painter, Paul Stevens and Jerry Taylor for their comments, suggestions and support.

Megan Black read through the entire first draft of the book and helped me improve it with many, many, many insightful observations on everything ranging from the basic structure of the argument to the subtleties of my theoretical position. I could not have wished for a better first reader.

I am grateful to Lester Crook, Tomasz Hoskins and Allison Walker of I.B.Tauris, who agreed to publish my work and then expertly guided me through the process of making a book out of a rather unfinished manuscript. I would also like to thank Ian McDonald, who patiently edited and polished the text until it read like proper English.

And then, finally, thanks to my two favorite people, Nina and Emma, for everything.

INTRODUCTION

A CULTURE OF DEPENDENCY

When faced with big energy challenges, the United States has "only two modes—complacency and panic". This is how James Schlesinger, who served as the first US Secretary of Energy in the late 1970s, later summarized the nation's basic approach to energy.[1] What Schlesinger's oft-quoted line suggests is that there is something profoundly emotional, erratic, even irrational in the way in which the American public and their policy makers have been dealing with issues in the field of energy.

US dependence on foreign oil, the main subject of this book, is no exception in this regard. The United States became a net importer of oil shortly after World War II, but throughout the postwar years oil imports remained a political issue confined to narrow policy and industry circles. This changed very abruptly in 1973, when a number of Middle Eastern oil producers implemented an oil embargo against the United States. In the midst of the energy crisis, the country shifted from relative complacency to near panic. President Richard Nixon had to explain to an angry American public that "we had allowed ourselves to become overly dependent upon foreign supplies of a vital good" and that, as a consequence, "the acts of foreign rulers, even far short of military action, could plunge us into an authentic crisis".[2]

Ever since the days of the energy crisis, oil imports have been an issue of high politics and public emotions in the United States. Every single US president since Nixon has described oil imports as a serious threat to national security, against which the country urgently needed to secure itself—for example, by achieving some form of energy independence, by changing its oil suppliers or by promoting alternative energy sources.

Despite such declarations, oil imports remained a significant part of the US energy system. On the eve of the embargo in 1973, about 35 percent of the petroleum consumed in the United States was shipped in from elsewhere in the world. After many ups and downs, and a record 60 percent in 2005, US import reliance most recently declined to about 33 percent of domestic oil consumption.[3] Besides regular presidential declarations, innumerable warnings about the dangers of foreign oil have also been issued by policy makers from both ends of the political spectrum; energy and security experts; researchers, soldiers and diplomats; writers, thinkers and journalists; and—in the form of opinion polls and surveys—by the American public at large. In fact, to say that US dependence on foreign oil presents the nation with serious problems of economic and national security would be considered a truism in most quarters in the United States.

Political discourse in the United States has been so saturated with references to the dangers of foreign oil and the vision of an energy-independent nation that some energy professionals have diagnosed the United States with a severe "obsession" about these issues. Two authors associated with the Cato Institute, for example, claim that worrying about foreign oil dependence has long belonged to "the most fashionable preoccupations in foreign policy circles". This is unfortunate, they say, because political concerns over producer blackmail, cut-offs and embargoes "are only a bit less far-fetched than worries about alien invasion". The notion that foreign producers may use oil as an economic weapon to hurt the United States is a narrative "fundamentally at odds with observable reality".[4]

While this is a particularly damning criticism of US energy security debates, the basic observation that something is not quite right with the way in which Americans have imagined the country's dependence on foreign oil is shared by many others. One energy analyst observes an "inane obsession with the idea of energy independence" in the United States.[5] Another describes anxieties over foreign oil as one of the country's most enduring policy obsessions, a complex revolving around deeply entrenched "mythologies of energy independence and foreign oil". The problem, as he sees it, is that "while emotionally compelling, these concepts are vestiges of a world that no longer exists".[6] The notion of an oil mythology also appears in the writings of Leonardo Maugeri, a well-known oil industry expert, who argues that oil has always "given rise to myths and obsessions, fears and misperceptions of reality". Among those, he counts a Western obsession with oil from the Persian Gulf, which he labels the "myth of oil security". Maugeri argues that claims about the national security implications of importing oil are mostly overstated. Even more worrisome, the policies to deal with this issue result from "a dangerous and even naïve miscalculation, both in political and economic terms".[7] Similarly, in a paper called "The myth of energy dependence" energy analysts Gal Luft and Anne Korin claim that US energy security debates are based on a historical misreading of the 1973 Arab oil embargo. Contrary to popular belief, they write, the United States never really relied on Middle Eastern oil. They too argue that the myth of dependence has affected policy making. For four decades, "Washington's energy policy has been based on the faulty conclusion that the country could solve all its energy woes by reducing its reliance on Middle Eastern oil."[8]

Claims about irrational anxieties, obsessions, myths, misperceptions and flawed understandings in US energy security debates are reflective of a certain perspective on energy issues that dominates the world of policy making as well as the majority of academic thinking about energy security in the United States. One of my principal goals with this book is to offer an alternative approach to understanding energy, and the question of dependency in particular. A good way of introducing this alternative perspective is to make explicit the basic, but usually unstated assumptions about the world that characterize the dominant view. There are two such assumptions. The first states that energy issues are an objective reality that can be grasped accurately by careful empirical study of the world. In this view, the challenge for those who think about US foreign oil dependence consists of assessing, as accurately and objectively as possible, the various costs and benefits of importing oil. What dependency is, what it means, how dangerous it is—all this exists independently of our own ideas and perspectives. It can be traced, measured, quantified and operationalized in terms of cause and effect. This kind of objective materialism in thinking about energy is perhaps most clearly visible in the various attempts to measure the "true" costs of US dependence on foreign oil.[9] But the same thinking also surfaces in the texts cited above, and those of many others who have identified oil-related myths and misperceptions in the US energy debate. Their argument is that there has been a divide between how Americans have imagined the dangers of foreign oil and a more sober, objective view of the issue based on studying the hard facts on the ground. A gap exists, as it is sometimes put, between "the myths and the realities of oil" or "between rhetoric and reality".[10]

Several energy writers, including some of those quoted above, have attempted to debunk the most common myths about foreign oil and energy independence.[11] And yet, despite many empirically valid and convincingly written attempts to reconcile energy debate and energy reality, many of the myths doggedly persist. As one policy analyst put it, America's present understanding of the world of oil still appears to be "frozen in the 1970s".[12] Even Leonardo Maugeri, whose book explicitly aims to explode the myths and obsessions of oil, has to admit that "neither explanations rooted in economics nor historical evidence have ever succeeded in countering the powerful mythology of oil insecurity".[13]

The second basic assumption underpinning the dominant way of thinking about energy is that issues in this field are clearly demarcated political problems that can and should be solved. In the United States, as Amitav Ghosh put it, oil is "a problem that can be written about only in the language of solutions".[14] Those who write about dependency, in effect, are almost always concerned in one way or another with policy making. Nothing symbolizes the dominance of this particular perspective better than the endless stream of think-tank studies, briefings, research papers, "roadmaps" and government reports on energy security and the various policy recommendations they have offered to decision makers in Washington over the past four decades. This genre of documents falls very firmly within the range of what Robert Cox describes as "problem-solving" approaches. Analysis of this kind, Cox explains, "takes the world as it finds it, with the prevailing social and power relationships and the institutions into which they are organized, as the given framework for action". The ultimate aim is "to make these relationships and institutions work smoothly by dealing effectively with particular sources of trouble".[15] Dependence on foreign oil is one such source of trouble, and thinking about the issue has focused on the question of how to make the United States secure against the dangers of foreign oil within the given political and economic parameters. In other words, the goal is usually to fix the problem of foreign oil. This is a pragmatic approach which, for reasons that are easy enough to understand, dominates the world of policy making.

A Cultural Perspective on Energy

This book suggests a different path to understanding energy politics.[16] At the heart of this perspective sits the suggestion that we should take American descriptions of dependency seriously in whatever form they confront us. Instead of assessing whether common understandings of foreign oil dependence represent an accurate description of an objective problem or insisting that they are irrational obsessions, myths, flawed interpretations of the world or empty political rhetoric, we could accept these views as a legitimate and important form of US political culture. Instead of attempting to bring rhetoric and reality into correspondence, we could see the stubborn persistence of fears about the perils of foreign oil as a sign that dependency has acquired significance for American society well above and beyond the narrow field of energy security. We could see debates about the meaning and consequences of foreign oil dependence as a site where Americans have formulated different views of what defines the essential nature of the nation, what the world is like and what America's place in it might be. In short, we could see dependency as a cultural discourse.

Reading the political debate about foreign oil dependence as a cultural phenomenon opens up an opportunity to study the cultural foundations of American energy policy since the 1970s, when the discourse of dependency first emerged. But this endeavor also yields a set of useful resources for making sense of the present. The country has been experiencing a spectacular boom in domestic production of unconventional oil and natural gas resources, which has begun to turn on its head much of the conventional wisdom about energy in the United States. The American energy revolution has breathed new life into the old debate about both US foreign oil dependence and the concomitant goal of energy independence.[17] These new developments, too, shape and are shaped by a pre-existing cultural discourse about energy and identity.

The following chapters present what may be described as a cultural history of oil in the twentieth and early twenty-first centuries. The focus will be on the issue of dependency as an object of politics: that is to say, dependency as an important political problem that compels the government to take action. Setting the scope of the book in this way, however, is complicated by the fact that American energy discourse has long contained two distinct, yet intimately related meanings for the term dependency. The first concerns reliance on imported oil; the second refers to the country's general dependence on oil as its major source of energy. While my main interest is in studying dependency as foreign oil dependence, it is impossible to ignore dependency as general oil dependence. Like sides of the same coin, the two concepts are inextricably linked. It will become clear early in the book that the very fact that different definitions of dependency have co-existed in a changing, ambiguous and sometimes contradictory relationship is in itself an essential part of the story. This is why the subtitle of the book reads *A Culture of Dependency* rather than *A Culture of Foreign Oil Dependence*. By default, when I speak about dependency on the following pages, I mean a discourse in which one or indeed both forms of oil reliance, foreign and general, are constituted as an important problem that requires policy makers to act. Where a clear distinction between different definitions of dependency is needed for the argument at hand, such a distinction will be given.

The cultural approach offered in this book departs from the conventional assumptions about how to consider energy in several important ways. It does not, to begin with, suggest a discussion of the "problem solving" kind. Instead of trying to formulate solutions to the problem of dependency or offering policy strategies to make the United States more energy secure, I am interested in a set of very different questions: what is the nature of dependency as an object of politics? What is this strange thing, US dependence on foreign oil, that has caused so much concern, anxiety and even panic over the past four decades? What can we learn about American society by examining the way in which this problem has been approached? This is not problem solving but much closer to what Robert Cox describes as "critical theory". In particular, asking these questions is in line with Cox's idea that a critical perspective should be "directed towards an appraisal of the very framework for action, or problematic, which problem-solving theory accepts as its parameters". Starting from one particular aspect or field, he explains, critical thinking is always oriented toward the entire social and political complex. It "leads towards the construction of a larger picture of the whole of which the initially contemplated part is just one component".[18] The element contemplated here is, of course, the problem of dependency; the larger social complex is American society.

The suggestion that dependency is a cultural discourse rests on two central arguments which, in one way or another, run through all the pages in this book. The first is that the meaning of dependency as a political problem is not simply inherent in the condition itself, nor does it emerge from an objective analysis devoid of social context. There is no necessary, simple or unproblematic correspondence between the thin, cold facts on the ground and the way in which Americans have imagined dependency. Such definitions have constantly changed. They always proved contingent, often unstable and sometimes contradictory. Historically, their emergence cannot be separated from the nature of the society in which they were forged. Dependency is a cultural construction.[19]

American culture and political definitions of energy issues are inextricably linked on a fundamental level. This argument is sometimes made, but usually only in a rather narrow sense. A number of authors, for instance, have mentioned culture as an important element in US concepts of energy security. Michael Klare argues that securing access to cheap and abundant oil is central "to the vigor and growth of the American economy and to the preservation of a distinctly *American* way of life".[20] In much the same way, Ian Rutledge argues that energy security in the United States is concerned first and foremost with preserving an American lifestyle based on oil-fuelled automobility. Effectively, energy security becomes about securing the American way of life. "National Security = Unimpeded Motorization."[21] These authors, in effect, draw a line between oil security and a very specific aspect of American culture: material culture, understood to encompass certain practices of energy consumption, lifestyle patterns associated with cars and far-flung suburbs, drive-in facilities and high levels of general per capita energy consumption. Bruce Beaubouef describes this view of energy security succinctly: "Policymakers found that protecting energy consumption patterns—in essence, protecting energy consumers—had become a vital part of national security."[22] Culture matters, the argument goes, because it is embedded in material practices of energy consumption, which then become the object of energy policy.

This is a good argument, but also one that can be taken much further. Practices of energy consumption do not constitute the only link between culture and energy, and are, arguably, not the most fundamental one. To say that dependency is a cultural discourse is to say that Americans have constructed its meaning out of a wide range of different cultural raw materials. These resources certainly include material elements such as patterns of energy consumption or path dependencies imposed by physical energy infrastructures, but they also include intangible social, political and linguistic elements that form part of American society: pre-existing American assumptions and narratives about energy and its role for human society that surface both in cultural and political contexts; an array of discourses which constitute objects said to define America's essential nature—notions like freedom, exceptionalism, independence, abundance or progress; collective American images of the world outside its borders and claims about the nature of the people who populate it; well-established political traditions and practices of representation in US foreign policy discourse; and established social relationships that assign to speakers in US society different degrees of political authority to explain the world. Showing how the meaning of dependency has been made and remade from these vastly different resources is one major ambition of the following chapters.

The second argument running through the book is that the discourse of dependency is linked to images of American national identity in complex ways. On the one hand, discourses about national identity and American images of the outside world are among the central cultural raw materials out of which political definitions of US foreign oil dependence have been articulated. As the following chapters will show, changing and conflicting political definitions of dependency, as well as struggles over how best to deal with the issue, have been closely bound up with varying claims about American society's relationship to energy and cultural images of the country's foreign suppliers of oil. In effect, studying dependency as a cultural phenomenon opens up an interesting vantage point from which we can examine evolving and fractured images of American identity, including discourses of Self and Other, in the United States.

On the other hand, the discourse of dependency has in itself served as an important means to validate and reaffirm certain constructions of national identity.[23] This is to say, Americans have used the discourses of dependency and energy independence to reaffirm who they are as a nation. Central to this process is the articulation by policy makers and others of a link between oil imports and serious dangers for the United States.[24] Besides threats to the national and economic security of the country, these dangers also include an existential threat to the United States as a society built on oil. The constant expression of these dangers, encapsulated also in the promise of regaining energy independence for the United States, has served to remind Americans about the founding values and principles that hold the nation together. In effect, the discourses of foreign oil dependence and energy independence have become an important platform for the maintenance and affirmation of American narratives of national identity.

In order to delve into the cultural discourse of dependency, we need to look at what various speakers in American society have had to say about the issue in the twentieth and twenty-first centuries. Again, the simple fact that the United States is importing oil from elsewhere is not inherently a serious problem of national security. Dependency has been constituted as an object of politics through practices of representation performed by concrete people in concrete social contexts: American policy makers, bureaucrats, soldiers, academics, journalists, oil company executives, industry lobbyists, environmentalists, advertisers, philosophers, novelists, film makers, cartoonists and ordinary Americans. What they have said, written or visualized about dependency provides a rich empirical basis for the following chapters. We can find in their documents articulations of the different meanings dependency has had in the United States, as well as traces of the cultural raw materials out of which these meanings have been formed.

Reflecting this broad approach to the subject, the historical materials discussed in the following chapters are very diverse. They include political speeches, internal and public documents produced by various federal government agencies, oil industry documents and advertising, polling data, academic and journalistic publications, energy statistics and other economic data as well as cartoons, novels, films and other works of fiction. That said, official documents produced by the federal government, in particular public presidential documents, are used more frequently than other sources. This choice reflects the observation that representatives of the State, and above all the president of the United States, are vested with special authority to define the national interest and give meaning to events and issues of foreign policy.[25] Sometimes, these official representations achieved the status of unquestioned truth. At other times, however, Americans opposed, challenged and contradicted the official narrative, and, with varying degrees of success, offered alternative readings of dependency instead. In both cases, presidential statements offer an excellent starting point for a broader examination.

A final remark about the scope of this book: it adopts a critical perspective on energy because it shares an interest with critical theorists in questioning the framework within which a certain problem has been dealt with. Instead of trying to fix the problem as it presents itself, sometimes it is more valuable to show that what is perceived as normal, self-evident or objective is really subjective, cultural, contingent and a little bit arbitrary. This, I would contend, is the case here. As a non-American, it is perhaps a little easier for me to see dependency as a profoundly cultural phenomenon and a specifically American experience. My main ambition is to show this through a detailed empirical study about the complex and fundamental links between the discourse of dependency and American society at large. For this reason, the book is perhaps best understood as a cultural history of dependency in the United States. There are, however, no more far-reaching critical goals. Specifically, I do not aim to find fault with the content of US definitions of dependency, I am not criticizing the social power relationships in which they have emerged, nor is it my intention to suggest a different, supposedly better way of understanding the issue.

Plan of the Book

Chapter 1, "Constructing the Hydrocarbon Society", provides an empirical foundation for the argument that political definitions of energy issues are cultural constructs. Borrowing a term from Daniel Yergin, the chapter argues that the United States is a "hydrocarbon society" in which oil has not only gained central importance for the economy and everyday life but also acquired an important position in cultural discourses about society and national identity. Focusing on the period between the 1920s and 1970s, when the key tenets of this society emerged, the chapter presents four themes with particular relevance for political debates about energy. Firstly, the discourse of abundance, which claimed that the country was richly blessed with natural resources and that American ingenuity and the power of the free market would make US oil reserves essentially limitless. Coupled with the belief that increasing energy consumption was causally linked to material progress and social mobility, the discourse of abundance provided a cultural foundation for the big energy choices shaping the hydrocarbon society. Secondly, oil became an important element in the American people's understanding of their basic rights and freedoms. In the context of the Cold War, oil and the free market were said to have provided Americans with free consumer choices and a level of material prosperity unknown in the Soviet Union. Thirdly, the central role of the oil-fuelled car in US society. As a technology, the car perfectly matched pre-existing American views about mobility, individualism and freedom. Through cars and driving, consuming oil became a practice central to the American way of life. Finally, the representation of oil as America's "lifeblood". This body-politic metaphor functioned as the primary linguistic symbol of America's relationship with oil. It suggested that oil was an internal, natural, benign, even healthy element of the nation's body, and that reliance on it was therefore unproblematic.

Chapter 2, "Defining Dependency", traces how these cultural raw materials have historically been used to assign meaning to America's reliance on oil. Written as a genealogy, the chapter explains how present political understandings of dependency have become logically possible. The energy crisis of the early 1970s was an important watershed in this process. Prior to the 1970s, dependency was not widely seen as a political problem. The energy crisis, however, rocked the cultural foundations of the hydrocarbon society—above all, the discourse of abundance—and enabled the emergence of dependency as a political problem. However, and this is key, two conflicting readings of dependency were formulated, which then struggled for dominance in the political debate. The first view, championed mainly by progressive segments of the public and the emerging environmental movement, defined the problem of dependency as American reliance on oil at large, and suggested a transition away from oil as the best solution. This definition emerged from a narrative of the energy crisis in which the problem had been caused by excessive domestic energy demand. Oil was not described as the country's lifeblood but as a drug to which Americans had become addicted. The second view defined the problem of dependency as American reliance on *foreign* oil and suggested a policy of energy independence as the best solution. This was based on a different narrative of the crisis, offered first by the US Government, which emphasized that oil had been withheld from innocent American consumers by a group of reckless Middle Eastern oil producers for purely political reasons. After the crisis, the latter reading came to dominate US energy debates. The alternative reading of dependency as general reliance on oil remained relatively marginal and only re-emerged in the 1990s. In the context of current debates about resource scarcities and climate change, this broader definition of dependency is today increasingly challenging the traditional equation of dependency with reliance on foreign oil.

Chapter 3, "The Dangers of Foreign Oil", examines the threats associated with foreign oil dependence. It argues that oil imports should not be considered a national security issue in the classical sense of military security or state survival, but instead as a threat to the viability of American descriptions of national identity. These threats were forged during the energy crisis of the 1970s. On the one hand, reliance on oil imports was understood in public and policy circles as a threat to the American way of life, both with regard to everyday patterns of energy consumption and with regard to cultural narratives of energy, economic growth, individual opportunity and social mobility. Secondly, dependency also appeared incompatible with the essential nature of the United States as an exceptional nation, a status reflected in the country's political independence, supreme power and global leadership. Policy makers, experts and the media argued that increasing dependence on foreign oil put the country "at the mercy of the Arabs", such that the concomitant loss of US power threatened not only the United States but the entire edifice of international order. These links between foreign oil and danger were formulated in the specific political context of the 1970s, but they persist and still shape the discourse of dependency today. However, new threats have also been identified. Most importantly, after 11 September 2001, foreign oil and the issue of international terror became entangled in American policy discussions about the Middle East. As a driver of US foreign policy, foreign oil dependence has also been described as a threat to the moral underpinnings of US activities in the world. The need to ensure oil supply meant that American policies had to conform to bare material necessities, even if that meant, for example, supporting hostile regimes in the Middle East.

Chapter 4, "The Foreignness of Foreign Oil", examines how American cultural images of the outside world have shaped political definitions of foreign oil dependence. American policy debates have long contained two radically different interpretations of oil imports.

Reliance on oil from some sources, above all the Persian Gulf producers, is described along the lines discussed in Chapters 2 and 3: as a dangerous threat to national security. Yet, imports from other countries, Canada in particular, are seen in the liberal tradition as a normal and mutually beneficial economic activity. This chapter asks: what determines the dangerousness of oil from a given foreign source? Focusing on the Middle East, it shows that the meaning of dependency is bound up with American cultural images of the places that provide the United States with oil. In this view, Canadian oil is not really foreign in any meaningful sense. Oil from the Middle East, by contrast, has been rendered foreign and dangerous by the region's long-standing status as one of the United States' principal cultural Others. In the 1970s, established American representations of the Arab world as a pre-modern, irrational and hostile place suggested that the region could not be depended upon to deliver the oil the United States increasingly needed. Moreover, the oil embargo was interpreted by many policy makers as a sign that the Arab producers rejected the central principles of liberal market economics as a basis for energy trade. They were said to have stepped outside the US-led liberal international order. After the 1970s, the region's cultural foreignness increased further. In US foreign policy discourse, the oil resources of the Middle East became entangled with Islamic extremism and terror, and they were sometimes inserted into a "clash of civilizations" discourse that emphasized the region's fundamental opposition to the founding values and principles of the United States. In these cultural contexts, a more benign reading of US reliance on Middle Eastern oil did not emerge.

Chapter 5, "Fighting for Oil", examines how different American descriptions of national identity and dependency have opened or foreclosed the possibility of using military force to safeguard US oil interests. After 1973, the US Government considered a military response to the oil embargo and the pricing policies pursued by the Organization of Petroleum Exporting Countries (OPEC). Calls for military action also emanated from the emerging neo-conservative camp. These proposals triggered a fierce public debate about the legitimacy of military action to safeguard national interests in foreign oil, a debate in which conflicting readings of the energy crisis surfaced and clashed. The majority of Americans rejected calls for war over oil as an illegitimate policy inconsistent with the nation's basic values. This view implied a striking critique of the assumption that oil was the nation's benign and natural lifeblood. Oil also played a major role in public discussions before the wars against Iraq in 1991 and 2003. In each case, official narratives emphasized the morality of war against widespread suspicions that private and national oil interests were really behind it. The ensuing debates again led Americans to consider whether war to safeguard the hydrocarbon society's constitutive raw material could ever be legitimate. The majority of Americans said no. In 1990, the public reacted angrily to official claims that safeguarding American national interests in oil was a legitimate reason for war. The slogan "no blood for oil" became the principal icon of opposition. The debate in 2003 also showed that oil was increasingly described as a dirty motif of US foreign policy. Oil became something of a political taboo, a dark subject linked to radical critiques of the government and to conspiracy theory. In a society that increasingly defined its relationship to oil as an addiction, there was less and less room for consideration of oil as a legitimate national interest.

CHAPTER 1

CONSTRUCTING THE HYDROCARBON SOCIETY

On the evening of 13 April 1949, the Hollywood motion picture *Tulsa* premiered in the town that carried its name. Starring Susan Hayward and Robert Preston, the film took its audience back to the great Oklahoma oil boom of the 1920s. It told the story of Cherokee Lansing, a strong-willed half-Indian farm girl, fighting her way up to fabulous wealth and power with the help of oil she finds on her family's land.

Tulsa did not do particularly well at the box office at the time. And yet, while it may not have been commercially successful, the film certainly is a very interesting cultural text about the role of oil in American life. *Tulsa* can teach us a great deal about the United States as an oil society, containing as it does almost all the major tropes in American oil discourse. There is oil as "black gold", creating opportunities for prosperity and social mobility for anyone in the United States. Then there is the corrupting influence of petroleum encapsulated in images of "Big Oil's" power and greed. Even the good-natured Cherokee Lansing finds it hard to resist the temptations of money and power. Importantly, there is also the destructive power of large-scale commercial oil extraction, transforming pristine nature into a hellish landscape of industrial waste. It is fair to say that *Tulsa* is an early example of environmentalist thinking about some of the inherent tensions between practices of economic production and the necessity to conserve the natural environment. As such, the film offers a stark reminder that current environmental debates about tar sands, offshore drilling or hydraulic fracturing for shale oil and gas have long historical roots.

Finally, *Tulsa* offers a very explicit narrative about the role of oil in modern American society. More than anything else, it is this tale of oil in American life that makes the film so interesting here. Oil is hailed as a catalyst for material progress, as an indispensable raw material for the American way of life, even as the very substance of modernity itself. In the opening scene, which is set in the present of the late 1940s, a folksy Oklahoma singer played by Chill Wills aims to give us some context for the ensuing story with a brief explanation of oil's emergence in American society:

> An oil well, a gusher blowing high and wide and handsome in the great state of Oklahoma. Oklahoma means red man's land, and less than fifty years ago, all this was Indian Territory [. . .] [They] raced their horses over the prairie, fished the streams, grew their crops and raised their cattle. And all the time the oil was underneath the ground. Well, it had to come out, and refineries had to be built to process it and make it what it is today: the lifeblood of our civilization. It's oil that drives the ships, powers the trains, and the planes, and fills the traffic lanes, and serves men in a thousand and one ways. Yes sir, oil's a mighty valuable commodity, sought for and fought for all over the globe.[1]

There are three points in these lines that nicely sum up the main ideas discussed in this chapter. First of all, the claim that oil had become America's "lifeblood" underlines the centrality of petroleum for all aspects of life in the United States. In 1950, soon after *Tulsa* was produced, oil accounted for about 38 percent of all primary energy consumption in the United States, overtaking coal for the first time. The United States, as observers were pointing out, had entered an "age of oil". Not only was the transportation sector almost entirely dependent on petroleum-based fuels, the petrochemical industry had also begun to develop a broad range of products that permeated US markets and soon became indispensable to consumers and companies alike—myriad forms of plastics and chemicals, lubricants, cosmetics, drugs, fertilizers and pesticides. Oil, as a 1945 industry publication put it, was America's "molecular treasure chest."[2]

By the end of the 1940s, the United States had become what Daniel Yergin aptly calls a "hydrocarbon society"—a society so fundamentally premised on the supply of oil that, as he writes, "we hardly stop to comprehend its pervasive significance". The notion of the hydrocarbon society is useful because it can take us beyond the simple observation that oil has been the most important source of energy in the US energy system. Oil does not just keep the wheels of industry turning. It also shapes, Yergin suggests, "where we live, how we live, how we commute to work, how we travel—even where we conduct our courtships".[3] In other words, oil has seeped into American culture. The hydrocarbon society is a vast socioeconomic structure that is made up of material, economic and technological elements but also an intangible yet dense fabric of symbols, practices and representations. It is a system based on oil-fuelled automobility, constituted by specific patterns of production and consumption, certain forms of social organization and by specifically American discourses about energy that give meaning to these practices and hold the entire structure together.

The second point concerns the origins of this society. As the narrator in *Tulsa* explains, a tremendous amount of human labor was required to make oil work in the United States. "Refineries had to be built to process it and make it what it is today." As the title of this chapter suggests, the hydrocarbon society had to be constructed. Building it required the redesign and partial removal of an older socioeconomic structure based on coal and steam and railways, which had itself been constructed as part of the Industrial Revolution. In order for the hydrocarbon society to come into existence, pipelines and other transportation infrastructure had to be put in place, oil technologies—the thermal cracking of hydrocarbon molecules, the oil lamp, rotary drilling, the internal combustion engine and many others—had to be researched, developed, manufactured, advertised and distributed on markets. Massive infrastructure projects to build thousands of miles of paved roads and highways required planning, legislation, funding and execution. A dense network of gas stations, garages, parking lots and retail stores had to be established. With every mile of freshly constructed road, with each new oil-based consumer product, the hydrocarbon society became a little bit more real. Americans had begun to construct the hydrocarbon society early in the twentieth century. Shortly after World War II, an energy system fundamentally based on oil had become a physical, manifest reality in the United States. It is still so today.

Thirdly, constructing the hydrocarbon society involved more than simply the development of certain technologies and physical infrastructure. Oil is a social construction. The minute it first gushed up from American soil in the nineteenth century, oil became an object of the American imagination. Laying the material base for the hydrocarbon society went hand in hand with a process of assigning meaning to oil as an increasingly important commodity, a process of imagining, representing, interpreting and narrating the energy-related change that the country was undergoing. Cultural texts like *Tulsa* demonstrate this clearly.[4] The lifeblood metaphor used in its opening scene offers one poignant example of how Americans actively sought to express the meaning oil had acquired for their society as a whole. The central point is this: constructing the hydrocarbon society was a process in which material and ideational practices were inextricably bound up with each other. The remainder of this chapter examines several key tenets of this society as they emerged between the early twentieth century and the energy crises of the 1970s.

Energy Eras and Energy Policy

In the United States, the history of energy is often written with the help of the twin concepts of "energy eras" and "energy transitions". Martin Melosi explains: "The concept of 'energy transitions' is based on the notion that a single energy source, or group of related sources, dominated the market during a particular period or era, eventually to be challenged and then replaced by another major source or sources."[5] American energy historiography usually observes two such transitions: the first one from wood to coal in the mid-nineteenth century, the second one from coal to oil in the first half of the twentieth century.

From a bird's eye view of aggregate energy statistics, oil appeared on the scene with the construction of the first oil well in Titusville, Pennsylvania in 1859. The following year, oil accounted for 0.1 percent of American energy consumption. By 1880, it had almost reached 2 percent, surpassed 6 percent in 1910, exploded to 24 percent by 1930, verged on 30 percent just before World War II and finally emerged triumphant at its all-time high of 44 percent in 1960. There it lingered for several decades, only to decline slightly in the 1990s and 2000s. In 2013, oil satisfied about 36 percent of America's energy hunger, still more than any other energy source. The use of coal, in turn, declined steadily after about 1910. Put most simply, the hydrocarbon society is the energy era resulting from this second energy transition.[6]

But these figures alone cannot give us more than a rough periodization of energy history. They say nothing about how these transitions happened and in what ways people used new energy sources to alter life in the United States. In order to capture the essence of this chapter's main argument—that the hydrocarbon society is both a physical and cultural construction—we need an approach to energy eras that is open to studying links between energy, technology and culture.

The work of David Nye contains many important ideas on which such an approach can be based. In his fascinating study *Consuming Power*, Nye examines historical linkages between energy technologies and culture in the United States. Nye does not see technology as an autonomous force that causes change in a society, but instead describes the concept of energy transitions as the "social construction of technological systems". Energy technologies, he argues, do not exist in isolation from the society in which they are created and used. American culture plays a role in selecting which technologies are used. These technologies then become part of the American lifeworld. "Each energy system", Nye explains, "has been used to shape distinctive domestic patterns, work routines, urban structures, and agricultural methods, imparting particular rhythms and contours to the everyday round." Energy, in other words, enters American society on the most basic level: it becomes part of daily life and thus helps define what people experience as normal.[7]

Beyond people's daily routines, energy is also part of culture in a deeper, less tangible way. As Nye explains in a different study, American culture is saturated with different "energy narratives", that is, widely held implicit or explicit conceptions about energy and its role in society, which surface in public debates, political arguments and cultural texts.[8] Through such narratives and other forms of symbolic representation, energy technologies can acquire, as Nye writes, "a dominant cultural position for decades or even generations".[9]

Nye argues that there is nothing inevitable about the rise of certain energy technologies over time. But he also emphasizes the importance of path dependencies in shaping the course of energy history. In his view, energy eras tend to persist for a long time because of their inherent "technological momentum". The emergence of any one energy technology may not be predetermined, but once the required physical infrastructure exists and becomes entrenched, it becomes very costly for a society to change course. As Nye puts it, "A society may choose to emphasize either mass transit or private automobiles, but a generation later it is difficult to undo such a decision."[10]

Nye's approach can be extended in a number of ways to provide a useful conceptual framework for the principal argument of this book. Most importantly, his contextualist approach to energy can be enlarged to encompass a more decisively political dimension. In other words, we can use Nye's social history of energy approach to write a social history of energy policy, based on the proposition that the process of formulating political definitions, goals and strategies in the field of energy is inextricably intertwined with the socially constructed energy eras in which these choices are made. Adopting such a perspective means, in essence, treating culture as an important factor shaping major energy policy choices. The argument here is not that energy culture simply determines energy policy as a causal force, but the rather different assertion that any dominant energy discourse makes certain political courses of action more urgent, logical, reasonable or justified while actively foreclosing other possible ways of acting. To write about energy in this manner is to enquire about the cultural conditions of possibility for the making of energy policy in the United States.[11]

In order to do this, we first need a better sense of the many ways in which oil has entered American culture in the hydrocarbon society—in the form of technologies, practices linked to energy consumption, energy narratives and other forms of symbolic representation. This is what the present chapter seeks to do. Only then can we examine how the political problem of dependency has been defined out of these cultural raw materials. This will be the aim of the following chapters.

One way of conceptualizing these links between energy culture and energy politics is to broaden the notion of technological momentum as offered by Nye. The argument that energy choices are made in the context of what is already there does not need to be confined to path dependencies arising from physical energy infrastructures. The hydrocarbon society also rests on a cultural energy infrastructure, an infrastructure that allows for certain courses of action more readily than for others. The concept of the "discursive economy" provides a useful tool for understanding how this works. David Campbell offers a working definition, according to which a discursive economy is "a managed space" in which "investments have been made in certain representations; dividends can be drawn by those interests that have made the investments; representations are taxed when they confront new and ambiguous circumstances".[12] Adapting the concept for the purpose at hand, we could say that the totality of American cultural practices, images, assumptions and representations regarding oil are stored in a discursive economy of oil. Some of these representations dominate in the sense that they have come to be seen as self-evident, obvious or true. Others remain marginal. Any political course of action in the field of energy is inevitably assessed, interpreted, weighed and debated against the reservoir of meanings contained in this discursive economy. In the process, some energy choices are constituted as more necessary or desirable than others. An example: during much of the twentieth century, American patterns of individual energy use were underpinned by a body of assumptions according to which energy consumption was a symbol of progress, a prerequisite for social mobility and rising living standards, an American birthright and an expression of the fundamental freedoms enjoyed by all US citizens. These powerful assumptions help explain a certain historical tendency in the United States to define energy problems in terms of supply, not demand. Writing about US energy policy in the mid-twentieth century, for example, oil historian David Painter concludes: "The realization that US oil consumption threatened to outpace domestic reserves led to plans to assure access to foreign oil reserves. The alternative of reducing, or at least slowing, the growth of rapidly rising consumption was not seriously considered."[13] Moreover, whenever such policy proposals to manage, restrict or reduce per capita energy consumption were made by policy makers—for instance in the wake of the energy crisis of the early 1970s—they had to fight an uphill struggle against established views about the benefits and necessity of consumption. It has usually proved difficult to act in ways which contradict the key cultural tenets of the hydrocarbon society.

American Oil Discourse

Oil has entered the American imagination in the form of many different narratives, metaphors and other forms of representation. The very material that is petroleum has been transformed by the alchemy of

American imagination into "black gold". The expression presumably dates back to the oil crazes of the nineteenth century and was perhaps meant as a reference to the early gold rushes in the American West. In the form of black gold, oil has functioned as an important symbol of the American Dream and the promise of upward social mobility. Oil could be struck on anyone's land at any time, bringing with it prosperity beyond imagination. Rags-to-riches stories of fabulous oil wealth are a staple in US popular culture. Besides Cherokee Lansing's story told in *Tulsa*, Upton Sinclair's novel *Oil!* comes to mind. The wealth of its main character, J. Arnold Ross, represents the real oil fortunes of the likes of the Rockefeller, Getty and Mellon families. Other well-known examples include James Dean as Jett Rink in the 1956 classic movie *Giant* or the popular 1960s and 1970s TV show *Beverly Hillbillies*, to name just a few.

The representation of oil as gold is also closely connected to the image of the oil man as a man working on the frontier: a rough, self-reliant American man operating at the margins of law and civilization; a twentieth-century cowboy. After the "real" frontier had closed with the end of westward expansion in the late nineteenth century, the geological rock formations between the surface of the land and the oilfields beneath it became a new frontier that these men could conquer. More than that, the oil frontier *had* to be conquered or oil would not have become a valuable source of energy in the first place. Reconsider the quote from *Tulsa's* opening scene. Oil had been there all the time, only the primitive Indians never discovered how to make use of it. It required true American ingenuity and manhood to bring oil across the frontier and make it part of civilization. In a similar fashion, American oil men would later conquer the oil frontier in other parts of the world. After 1945, for example, Saudi Arabia and its oilfields became a new "American frontier".[14] Oil companies actively promoted this image of themselves. In 1953, to take one example, the American Petroleum Institute sponsored a short documentary film called *American Frontier* about oil discoveries in North Dakota. The key message was simple: oil brings progress and wealth to local communities and the nation as a whole, but only if brave pioneers first fight the harsh land for the treasure buried beneath it.[15]

In stark contrast to these rather positive representations of oil and oil men, and somewhat paradoxically, the oil industry itself has had an extremely bad reputation in the United States almost from day one. The history of these negative images can be traced back to the rise to power of John D. Rockefeller's Standard Oil Company in the late nineteenth century. By 1880, the company controlled over 90 percent of refinery capacity in the United States. Standard Oil became, in the words of Roger Olien and Diana Davids Olien, the "tangible and cogent model" of the industrial monopoly in the United States. The company, and with it the entire petroleum industry, became subject to a widely held body of negative assumptions. "According to this construction, the industry was monopolistic, overpowerful, speculative and risky, conspiratorial, wasteful, disorderly, out to gouge consumers, out to corrupt government, and, in general, a threat to public welfare."[16] Americans still think of Big Oil in similar terms today. Public outrage and conspiracy theories about "obscene profits" and government—industry collusion have mushroomed during every energy crisis in hydrocarbon America.[17] Naturally, the oil industry has tried to change its image through various public relations measures. A fine example of these efforts was Mobil Oil's weekly "Op-Ed" series of editorial-style ads in the *New York Times* and other major newspapers during the 1970s, a campaign explicitly aimed at strengthening "Mobil's reputation as a socially responsible corporation."[18] However, the American public has remained unconvinced to this day. Opinion polls show that, over the years, Americans have consistently rated the oil and gas industry worse than almost any other business sector in the country.[19]

These three themes—oil as black gold, oil men as frontier men and oil companies as a threat to public welfare—are among the foundational elements in American oil culture. Others include the representation of oil as an environmental threat or the articulation of an intrinsic connection between oil and blood. Some of them will be discussed later in the book. The principal aim of this chapter, however, is to study those cultural raw materials that have been most influential in shaping political definitions of dependency in the United States: firstly, the assumption that there is a connection between energy, abundance and progress; secondly, the assumption of a connection between oil, capitalism and freedom; thirdly, the role of oil and the American system of automobility; finally, oil and American self-representation—that is to say, the way in which people have imagined the significance of oil for their own lives.

Progress, Abundance and the High-Energy Society

If there is one claim that is foundational to the cultural construction of the hydrocarbon society in the twentieth century, it is the notion that consuming more energy is a good thing. We might dismiss this assumption as absurd from today's perspective, but the articulation of increasing energy consumption as a precondition for, and measure of, the progress of human civilization has a long and powerful history. In the 1930s, the American archaeologist G. G. MacCurdy formulated this claim in straightforward terms: "The degree of civilization of any epoch, people, or group of peoples, is measured by ability to utilize energy for human advancements or needs."[20] More energy automatically meant more progress. This view of energy consumption is still present today. Consider this example from a recent book about energy in the United States: "Energy consumption creates wealth. It is axiomatic: As energy use rises, people get richer. It's true always, everywhere."[21] In this manner, energy has been linked to economic growth as well as material and cultural progress, and this connection has sometimes been raised to the status of a law-like causal force in the grand equation of human civilization. Macroeconomic forecasting, to take another example, has long been based on a so-called iron law of energy—the notion that economic growth necessarily correlates with a proportionate increase in energy use.[22] Few people would make such bold claims today. But at the time of the energy transition that brought forth the hydrocarbon society in the United States, the articulation of a causal link between energy consumption and progress dominated energy discourse. In her analysis of twentieth-century economics textbooks, Stefania Barca shows that increasing use of inanimate energy sources was routinely represented as an "undisputable accomplishment" of society, as a measure of humanity's "liberation" from the shackles of its material condition.[23] Similarly, oil consumption in the United States was routinely described as an inherently positive activity. As Daniel Yergin writes, "For most of the twentieth century, growing reliance on petroleum was almost universally celebrated as a good, a symbol of human progress."[24] How did such views contribute to the emergence of the hydrocarbon society? They did so, for example, by providing space for the argument that increasing energy consumption should be an explicit goal of US energy policy. Again, this is an idea that appears strange from today's perspective, but according to Amory Lovins, most

US energy policy proposals up until the 1970s promoted the goal of "sustaining growth in energy consumption" as this was assumed to be causally linked to continued economic growth.[25]

In the United States, the assumption of a link between energy use and human progress is entangled with a cultural discourse of abundance. In Europe, where energy has historically been scarce, there is no such discourse. But in the United States, as David Nye points out, "for 150 years actual scarcity had seldom been a concern when an energy choice was made".[26] The transition from coal to oil, for instance, was by no means the result of the looming depletion of coal reserves. If there were energy-related problems in the history of the United States, Martin Melosi agrees, they were "problems of choice" and the resulting necessity to somehow "cope with abundance".[27] It is important to note that the hydrocarbon society was constructed in a situation in which all major sources of energy—coal, oil, natural gas—were available in abundance. The historical experience of natural resource wealth, Nye continues, contributed significantly to the emergence of a widespread sentiment in the United States that abundance was natural while shortages and scarcity were artificial.[28] Abundance had become, as David Potter observed in the 1950s, a veritable social factor in the United States: "For certainly it is an influence that impinges upon all American social conditions and contributes in the most fundamental way to the shaping of the American culture and the American character."[29]

The representation of abundance as the natural state of affairs in the United States culminated in the claim that US oil supply was essentially without limit. The US oil industry actively propagated this notion of limitlessness. In 1952, Eugene Holman, then CEO of Jersey Standard, published an article in the *Atlantic Monthly*, which he called "Our inexhaustible resources." Of course, Holman knew that there was a finite amount of oil under American soil but, as he argued, through scientific progress, better exploration and improved drilling technologies oil could be made limitless for all practical purposes:

> In the United States alone, there has been produced since 1938 as much oil as was known to exist in the country at that time. And despite that great withdrawal, the domestic industry's proved reserves are at an all-time high level. It's as though we started out with a tank of oil, used it all up, and had a bigger tankful left.

Holman concluded that America's oil reserves were so vast that an end to abundance was "beyond sight".[30]

After 1945, these representations of abundance, energy consumption and progress helped bring about what Melosi calls the "uncritical notion of an optimistic materialistic future" in the United States. They "fostered a belief in unbridled material progress linked to visions of a high-technology society, where every family drove a new car, had a modern ranch-style home, and accumulated every appliance that could be plugged into a wall socket".[31] Fuelled by cheap and abundant oil supply, Americans experienced the decades after the war as a time of spectacular growth, confidence and expansion on every level. "Bigger is better" was the catchphrase of the times. Between the 1930s and the 1970s, total American energy consumption exploded, almost unbelievably, by 350 percent. The United States became the most energy-intensive society in all of human history. In 1970, the average American household had more energy at its disposal than an entire small town in Colonial times. The country had truly become a "high-energy society".[32]

The American oil industry used these pre-existing themes for its own ends, and invested heavily in the dissemination of positive energy messages. After the war, the oil companies printed a range of educational brochures, leaflets, books and other materials to educate the public about issues regarding oil and US society. In 1945, for instance, Standard Oil of New Jersey produced a little book of 31 pages for education in schools. Teachers could order free copies which came with "teacher's manual" leaflets that provided guidelines about how best to convey the ideas in the book. In addition to providing basic information about the oil industry, this publication offered a narrative drawing on the connections between oil, progress, abundance and limitless energy for everyone:

> Without gasoline, we might still be in the horse-and-buggy age. The energy contained in our Nation's present output of oil equals the work done by 4,500,000,000 men working eight hours a day, six days a week. This is the same as 31 able-bodied servants working for each man, woman, and child in the United States.

Proudly, Jersey Standard announced that Americans were using 30 times as much oil as the average person in the world and that oil was "actually cheaper than dirt" in the United States. "We shall need more oil in the future. And there will be oil in plenty." According to this narrative, it was the country's abundance of oil that had helped it attain a standard of living that was the envy of the world. Abundance was "freeing men's hands and minds and spirits for opportunities to do more than earn mere existence."[33]

Consumption, Freedom and America's Cold War Identity

An important aspect of the energy transition from coal to oil in the first half of the twentieth century was that the products supplied by the petrochemical industry became increasingly important to American life. Timothy Mitchell argues that oil became a commodity central to the management of a new type of economy based on the notion of growth. Due to declining energy prices, oil's seeming inexhaustibility and the agricultural revolution it enabled, as well as its many industrial uses, "oil contributed to the new conception of the economy as an object that could grow without limit".[34]

Mass consumption played a key role in this growth-oriented postwar economy. The key tenets of consumerism are much older than even the Industrial Revolution, but in the twentieth century new market institutions revolutionized consumption practices in the United States. Professional advertising, department stores, mail-order catalogues and, some time later, shopping malls permitted the emergence of non-spatial consumption communities.[35] After 1945, as Lizabeth Cohen argues, there emerged a new political consensus in the United States that continuous economic growth was required to keep the promise of prosperity for all. Since consumption was a central pillar of economic growth, it was to be actively encouraged. "The purchaser as citizen", Cohen writes, "simultaneously fulfilled personal desire and civic obligation by consuming."[36]

The hydrocarbon society, in effect, has been both a high-energy and a mass-consumption society. Moreover, these two facets are closely connected, as mass consumption was premised on a readily available surplus of energy.[37] Cheap oil provided the fuels and lubricants that enabled the large-scale production and transportation of goods, while oil also provided the basis for countless forms of consumer goods that could be purchased by Americans. The best example of a product linking mass consumption and oil is, of course, the automobile. It is the quintessential consumer good, and will be discussed in some detail below. And yet, while the car is certainly central to understanding the hydrocarbon society it is only one product among a vast range of oil-based goods, electrical appliances and materials furnished by the manufacturing and petrochemical industries. The hydrocarbon society is not just a car society.

Mass consumption as a cultural practice could only flourish to the extent that simple purchase decisions were rendered meaningful within a discourse of consumption. Advertising provided a powerful tool to disseminate such messages. In the 1920s, to take just one example, General Electric worked tirelessly to instill an "electrical consciousness" in the minds of Americans. The company circulated up to 5 billion print advertisements per year in order to create demand for new electric appliances. Many of these products were overwhelmingly successful. After World War II, consumers wanted, and could increasingly afford, washing machines and refrigerators, vacuum cleaners and radios—and,

Figure 1.1 Men can make many things we need from oil, 1957. Companies used various means of advertising the endless consumption opportunities presented by oil. This book aimed to teach young children about oil as a fuel for cars and as the basis for many of the products that made life so good in the United States.[38]

a little later, also TV sets, air conditioners and computers. Advertisers told them that owning those products constituted the American way of life.[39]

Alternative ways of promoting the endless possibilities of consumption existed and were widely used. Consider, for example, a children's book produced by Esso in 1957, which aimed to introduce very young children to oil and its many uses. Just look how the children's faces are glowing with enchantment as they experience this miraculous diversity (Figure 1.1). Esso's message was simple: petroleum was so much more than just a foul-smelling black liquid; it was the basis for the things that constituted the good, modern life.

Through its role in practices of consumption, oil has been linked to constructions of American identity in the hydrocarbon society. Consumption has become a basic building block of individual personal identity.[40] More importantly, consumption is also linked to collective identity on the national level. After World War II, practices of consumption and the high living standards that they promised became entangled with two key constitutive elements of American identity: freedom and democracy. In the context of the emerging Cold War, American policy makers began to represent mass consumption as a symbol of individual freedom and democratic political rights. Material progress was seen as evidence of the superiority of the American system vis-à-vis the Soviet Union, which was described as a "slave state" whose entire purpose was to "eliminate the challenge of freedom".[41] The political freedom of Americans was echoed in their free choice as consumers and the abundance of material goods they encountered in their everyday lives. Consumption was becoming a kind of American birthright.[42] This was made explicit, for example, during the 1959 American National Exhibition in Moscow by the then Vice-president Richard Nixon. Together with Soviet Premier Nikita Khrushchev, Nixon visited a model American home, complete with modern electronic appliances and an American-made car. He then delivered a speech entitled "What Freedom Means to Us" in which he cited the unstoppable march of material progress as a key constituent of the American way of life. The United States, Nixon said, was as close as anyone had ever come to the ideal of prosperity for all in a classless society. He pointed out that practically every American could afford a home like the one displayed in the exhibition. In the ensuing impromptu argument between Nixon and

Khrushchev, known as the "kitchen debate", Nixon hailed the material abundance of the American economy as evidence for the superiority of the American free-market model.[43]

Abundance was an important theme in the American discourse of consumption. It was reflected in endless consumer choices that allowed Americans to exercise their basic freedoms. But at the same time, abundance was also articulated as an effect of freedom. Abundance was, as David Fischer puts it, both "the source and substance of American freedom".[44] After 1945, the oil industry made this point over and over again. The message was always the same: the United States had only a fraction of the world's petroleum reserves, yet no other country had achieved the same abundance in oil. The magic of the free market was behind this spectacular success. Consider the following speech by Jersey Standard's Eugene Holman, given in 1944:

> If we have not been especially blessed with oil by Nature, how does it happen that we have been so successful? The answer, I believe, is to be found in our form of government and society—in democracy [. . .] As a result, we have found so much oil that we have given America an industrial capacity, a motor transport, a standard of living for the mass of our people unequalled in any other nation [. . .] Above all, we must have freedom of competition. I have said that oil is a product of freedom. But wherever there is freedom of action, there is competition. In a sense, competition is a corollary of freedom, and oil is also a product of competition.[45]

Corporate efforts to disseminate these messages were by no means limited to lobbying the government. They were intended for a much wider American audience. In 1956, the American Petroleum Institute sponsored the production of an animated short film called *Destination Earth*, which was shown in cinemas before the main feature. The film provides us with a fine example of all the major themes in the American oil discourse discussed above: the achievement of abundance through freedom and competition, the role of oil in creating material prosperity and high standards of living, and the affirmation of American Cold War identity through the claim that the United States and the Soviet Union were radically different.

In *Destination Earth*, the red planet, Mars, represents the Soviet Union. It is ruled by a dictator who bears more than a passing resemblance to Joseph Stalin. The planet's centrally planned economy is in a shambles and, even worse, suffers from a critical energy shortage. In order to solve the problem of energy supply, the dictator sends space explorer Colonel Cosmic to Earth to discover the secret of human energy production. As it so happens, our hero lands in the United States where he soon encounters the automobile ("Great Ganymede, they were superb!"). To his awed surprise, he discovers that in the United States such luxuries are not available only to a chosen few but to nearly everyone. And even more importantly, he learns that it is oil that fuels the automobile and that access to oil in the United States is not regulated by the central government. In fact, thousands of companies compete to explore and drill for petroleum, transport and refine it, and then offer it to consumers: "A whole galaxy of things to make a better life on Earth!"

Back on Mars, our hero reports his findings to the planet's ruler, who is intrigued by oil but dismisses the free-market principle as "downright un-Martian". But it is already too late. The march of freedom is unstoppable. The people of Mars are instantly won over by the promise of free-market prosperity and the dictatorship is swept away as hundreds of entrepreneurial Martians go search for oil. To drive an already obvious point home, a narrator concludes the film by saying:

> The big secret is of course oil, which has brought a better life to all the people in the USA. But the key to making oil work for everybody is competition [...] Yes, the real secret is not only a great source of energy, but also the freedom to make it work for everybody. And if you have both of these things, any goal is possible. It's destination unlimited![46]

The Hydrocarbon Society as a Car Society

The car is the ultimate symbol of the hydrocarbon society. Not only is it the quintessential manufactured product and, as Melosi puts it, the "granddaddy of all consumer goods", but the gasoline-fuelled internal combustion engine has also been the single most important oil-related piece of technology in the United States.[47] Almost all forms of transportation in the country are dependent on oil today. The majority of oil consumed in the United States is used in the transportation sector. In turn, over 90 percent of fuels used for transportation are refined products based on crude oil. No wonder then that "the pump" has become an important place in American society.

The car should not be seen as an isolated piece of technology. What cars are, what they mean, and what they can do is enabled by a dense web of other technologies, physical infrastructure and cultural practices. This framework has been called the American "system of automobility". The centrality of the car for the hydrocarbon society, "its specific character of domination", as John Urry puts it, can only be understood in terms of this system. According to Urry, the car is the primary manufactured good, an important consumer article that provides social status to its owner, the most important form of personal mobility in the United States, a central object in American discourses about what constitutes the good life and a major driver of resource use and environmental degradation.[48] To a significant degree, constructing the hydrocarbon society amounted to constructing this system of automobility. These deep connections are echoed in the "symbiotic relationship" between the oil and car industries that developed after 1945, and it is not an exaggeration to say that, after the war, the United States became an economy based on oil and cars.[49] In 1960, General Motors, Standard Oil of New Jersey and the Ford Motor Company occupied the first three places on *Fortune Magazine's* list of the largest American corporations.[50]

The adoption of the oil-fuelled automobile as the nation's favorite means of transportation was a contingent social process. Initially, factors such as the sheer size of the nation, the lack of decent paved roads and relatively dense systems of mass transportation worked against the spread of cars. In the early 1900s, the internal combustion engine also faced significant competition from steamers and electric cars. It has been argued that the oil-fuelled car beat other forms of transportation, as well as competition from electric cars and steamers, because of free-market economics and sovereign consumer choices. In this view, the entrepreneurial spirit of Henry Ford, who founded a motor company that specialized in internal combustion engines, was essential to the rise of the automobile. It was the Fordist system of mass production that permitted the incredible success of the Model-T. Thanks to economies of scale, sales prices dropped with increasing manufacturing quantities. While in 1900 a mere 4,200 cars had been manufactured in the country, by 1929 this figure had already grown to 5.6 million. "Society decided it wanted the automobile, and it bought the automobile."[51]

Others have expressed doubts about such purely economic arguments, arguing that the emergence of the oil-fuelled automobile and the dismantling of America's once widespread network of public transportation should be seen as the deliberate act of a corporate petroleum—automobile complex.[52] These studies emphasize the role that powerful actors played in the construction of the system of automobility. Besides influence over economic and political decision making, however, there was also a more subtle form of power at work in the construction of this system—one in which speakers relied on pre-existing elements of American culture to give meaning to technologies like cars, or energy sources like oil. Corporations, in other words, worked to create not only the physical infrastructure that cars required but also a cultural infrastructure to make the car work in American society. Advertising became an important tool in the exercise of this kind of discursive power. The oil and automobile industries turned to mass advertising in the 1920s. Madison Avenue ad-man Bruce Barton lectured oil executives about how "the magic of gasoline" needed to be harnessed not only in the engines of their cars but also in terms of branding and advertising:

> Stand for an hour beside one of your filling stations. Talk to the people who come in to buy gas. Discover for yourself what magic a dollar's worth of gasoline a week has worked in their lives. My friends, it is the juice of the fountain of eternal youth that you are selling. It is health. It is comfort. It is success. And you have sold merely a bad smelling liquid at so many cents per gallon.[53]

In the face of falling real oil prices and an increasingly saturated automobile market, companies started to compete with each other for the attention of consumers. The big three car makers—Ford, General Motors and Chrysler—tried to gain market share by producing ever bigger and more powerful cars. The oil companies, for their part, started to brand gasoline and offer better service as well as free giveaways at their gas stations. Esso's successful 1950s slogan "put a tiger in your tank" is a well-known example for this strategy.[54] Corporations produced cars with in-built obsolescence, both in a material and an emotional sense. This ensured profits and kept the wheel of American mass consumption turning. As one Ford designer put it in the 1950s, "We design a car to make a man unhappy with his 1957 Ford 'long [sic] about the end of 1958."[55]

Corporate messages about the automobile worked so well because they could draw strength from pre-existing themes deeply embedded in American culture. For example, the availability of privately owned means of transportation fit into a powerful assumption about the importance of mobility in the United States. David Potter argues that Americans have always equated spatial mobility with the promise of social mobility. In the context of the frontier moving westward in the nineteenth century, mobility was synonymous with equality of chances, which in turn meant freedom.[56] In much the same way, observers of American life from Alexis de Tocqueville onwards have emphasized the restless and individualistic character of the American people. In 1905, Cambridge historian G. Lowes Dickinson stated that "to be always moving, and always moving faster" is what Americans "think is the beatific life".[57] Much more than the railroad or other forms of mass transportation, the automobile could be presented as a desirable object within this cultural framework. The car came to symbolize freedom on wheels. As one study describes it, "Automobility facilitates individual determination, individual freedom of movement, self-directedness, privacy, choice of destination arrival time, and control over immediate environment. To many, automobility is at the core of individualism in America."[58]

The car had become an essential part of the American way of life by the 1920s. As Melosi writes, cars achieved "the status of a social, if not an economic, necessity".[59] Moreover, the dense network of support infrastructure required to make these vehicles work—dealerships, garages, gas stations or specialized retail outlets—had been put in place for one type of car only: the gasoline-fuelled automobile. Steamers and electric cars soon disappeared from the market.[60] After World War II, car ownership increased rapidly with rising living standards. Registration numbers increased by 50 percent in just ten years, from 49 million in 1950 to 74 million in 1960.[61] Meanwhile, the relevance of public transportation declined. There had been about 17 billion passengers on buses and trolleys in 1950. By the 1970s, this figure had decreased to 7 billion per year. Hundreds of companies were forced out of business.[62] The United States had acquired a "one-dimensional transportation system".[63]

The physical infrastructure required to support private automobility was created at the same time. Thousands of miles of paved roads were built. As early as 1916, the concept of a national highway system was created. It took until the mid-century, however, before construction of this system really kicked off with the Interstate Highway Act of 1956. The bill provided the funding for the construction of 41,000 miles of toll-free express interstate highways.[64] By the 1970s, 3 million miles of paved roads existed in the United States—up from 378,000 miles in 1920.[65] This process went hand in hand with a large-scale reordering of the urban landscape. Suburban neighborhoods were designed on the premise of individual car ownership. According to one estimate, up to 50 percent of urban space had to be freed up for downtown parking lots, gas stations, traffic lights, car-friendly roads and other infrastructure in order to enable the car to function in the United States.[66]

The social changes that accompanied these developments have been described as "nothing short of revolutionary".[67] The emerging system of automobility permitted the spread of new forms of living and the development of a "drive-in society" in which Americans could eat, shop, pray, watch a movie, enroll on college courses or experience their natural environment without leaving their cars. Even more fundamentally, the automobile "became the vehicle of ideologies about the city, the home, and the world of consumption".[68] Cars became central to shaping American gender roles, signaling the beginning of adulthood and defining modes of sexuality and courtship. For generations of Americans, getting their first car became a rite of passage into adult life. In the late 1960s, according to Yergin, 40 percent of all marriage proposals took place in cars.[69] Through cars and driving, the private consumption of oil became an important part of everyday life in the United States, and an activity inextricably intertwined with the maintenance of the personal and social identities of modern Americans.

The Self-representation of the Hydrocarbon Society

Americans have always searched for ways of expressing their own relationship to oil, and these descriptions are essential to understanding the cultural foundations of energy policy in the United States. This process of self-representation started at the very beginning of the energy transition from coal to oil, and will never be completed as long as oil retains its position in the American economy. Studying these representations and how they have emerged, evolved and clashed with each other over time opens up a window through which we can look into the very core of the hydrocarbon society as a cultural construction.

Consider one more time the excerpt from *Tulsa's* opening scene with which this chapter began. Oil had to be discovered and refineries had to be built to make oil "what it is today: the lifeblood of our civilization". The metaphor of oil as America's lifeblood deserves attention. Between the early days of the hydrocarbon society, in the first third of the twentieth century, and the 1970s it was the most commonly used expression with which this society represented itself. In fact, it is fair to say that if the car is the material object that best symbolizes the hydrocarbon society, the lifeblood metaphor is its principal linguistic symbol. In American accounts of oil, irrespective of whether they are works of fiction or dry academic prose, the phrase is ubiquitous. Its genealogy can be traced back to the very beginnings of the hydrocarbon society. As early as 1933, Interior Secretary Harold Ickes stated during a Senate hearing that oil "may in time prove to be the very life-blood of the Nation".[70] What appeared as a mere possibility in this statement became a manifest reality after World War II. The lifeblood metaphor established itself as a standing expression, and became a staple in American oil discourse. Although, as the following chapter will show, the situation became more complicated in the 1970s and even more complicated in the most recent past, oil is still widely described as lifeblood today. Here is just one example from an energy report published in 2006: "Oil is the lifeblood of the American economy, providing more than 40 percent of all energy consumed in the United States and 97 percent of the energy used for transportation."[71]

It is no coincidence that the self-representation of the hydrocarbon society has taken the form of a metaphor. Metaphors are an important means by which people make sense of their natural and social environment. What is more, they are often used to express how something relates to society as a whole. Body-politic metaphors have been used by people across all times and cultures to give meaning to events, objects or places in socio-medical terms—that is to say, in relation to society imagined as a health human body. The lifeblood metaphor forms part of this tradition, conceptualizing American society as a human being and oil as one if its constitutive parts.[72]

It is worth questioning what it really means to say that oil is the lifeblood of the nation's body. There is a vast reservoir of meaning just underneath the surface of this metaphor. Studying these meanings reveals much about how Americans have imagined the role of oil in their society, including the question of dependency. For a start, representing oil as lifeblood suggests that oil is not only internal to the body politic but a natural, constitutive part of this body. The emergence of the lifeblood metaphor signified that oil had crossed the frontier and was now part and parcel of modern civilization. The technologies and infrastructure required to make oil work were in place. Secondly, the lifeblood metaphor implies that oil is essential to the health and very existence of the nation. Without oil the body politic could no longer function. While the lifeblood metaphor therefore contains an implicit admission of US dependence on oil, in no way does it imply that this reliance is inherently problematic and needs to be changed. More than that: reliance on oil cannot be avoided since oil is naturally a part of the American body politic. Dependency appears as a simple, inevitable fact of life. Hydrocarbon America relies on oil like the human body relies on oxygen: it cannot be any other way.

Conclusion

Upton Sinclair's novel *Oil!*, published in 1927, is written from the perspective of a young man called Bunny, son of self-made oil millionaire J. Arnold Ross. The early chapters of the book describe Bunny's childhood during the California oil boom of the early twentieth century. They provide a vivid account of the wonder Americans must have experienced while the hydrocarbon society was slowly emerging all around them. On one occasion Bunny is being driven across the mountains of southern California in his father's brand new car, a cruise he is thoroughly enjoying:

> Any boy will tell you that this is glorious. Whoopee! you bet! Sailing along up there close to the clouds, with an engine full of power, magically harnessed, subject to the faintest pressure from the ball of your foot. The power of ninety horses—think of that! [...] And this magic ribbon of concrete laid out for you, winding here and there [...] twisting, turning, tilting inward on the

> outside curves, tilting outward on the inside curves, so that you were always balanced, always safe—and with a white-painted line marking the centre, so that you always knew exactly where you had a right to be—what magic had done all this?[73]

One answer to Bunny's question is suggested by the title of Sinclair's book. This chapter offers a similar, yet more complex answer: the hydrocarbon society is the product of a complicated construction process that combined the creation of physical infrastructure, technologies and markets with the cultural construction, less visible but equally important, of oil as the country's dominant source of energy. It should be clear now that American energy culture contains a wide range of themes, narratives and representations, and that these are directly linked with issues that have had an important role in defining American identity in the twentieth century: abundance, consumption, prosperity, progress and freedom.

By the 1960s, the central tenets of the hydrocarbon society were firmly in place. Then, as Rutledge writes, "the die was cast. The whole socio-economic structure of American society [. . .] had been constructed in a manner which would make it virtually impossible to modify in future years without a major effort of political will."[74] Investments had been made in the American economy that gave the hydrocarbon society a technological momentum that still drives important economic developments today. These include the physical infrastructure to enable a mode of transportation based on the private automobile; to explore, drill for and refine oil; and the networks of distribution, retail outlets, malls, shopping centers, drive-in facilities and motels, as well as suburban modes of living, which still define much of American life in the twenty-first century, and which now appear to many as obstacles on the way to a different, presumably greener energy future.

But investments had also been made in a discursive economy of oil, which served as a catalyst for the emergence of the hydrocarbon society and then helped support and perpetuate it. These include narratives that equated oil with material progress, abundance, social mobility and the promise of individual opportunity, and which linked consumption to American Cold War identity and cars to the exercise of individual freedom. These discursive investments acquired power partially because they could be inserted into pre-existing assumptions about what constituted "the good life" and what it meant to be American. They acquired power also because actors with discursive authority, including corporations with vested interests in the construction of the hydrocarbon society, worked tirelessly to construct oil and oil-fuelled technologies as desirable and quintessentially American.

CHAPTER 2

DEFINING DEPENDENCY

George W. Bush produced many memorable lines during his two terms as president of the United States. Not many of his statements, however, surprised Americans more than the one he made on 31 January 2006, when, 36 minutes into his annual State of the Union address, Bush suddenly declared that "America is addicted to oil, which is often imported from unstable parts of the world." Calling dependency "a serious problem", the President went on to admonish his fellow Americans that they had to "break this addiction" by changing the way they heated their homes and powered their automobiles.[1]

The President had issued a declaration of dependence: the admission that something was fundamentally wrong with the country's reliance on oil. Political observers reacted with astonishment to Bush's speech. As the *New York Times* observed, "The president's tone was so changed, in fact, that some analysts said he sounded like a Democrat."[2] But the fact that Bush, a Texas Republican, had espoused the kind of rhetoric previously used mostly by environmentalists and progressive liberals was only one reason why Americans were surprised. No incumbent US president, Republican or Democrat, had ever publicly spoken about national reliance on oil in terms of an addiction. While the use of the addiction metaphor can be traced back to the energy crisis of the 1970s, it had not gained traction in the key policy circles in Washington before. Now, there was a sense that when the president of the United States made a statement like that in the strong symbolic context of the annual State of the Union address, this was bound to have an effect on the country at large. Thomas Friedman of the *New York Times* predicted,

"When the president changes language on an issue like this [. . .] the whole country and bureaucracy start to talk differently."[3]

Friedman was right. In the wake of Bush's speech, the notion of an American oil addiction became an instant catchphrase. It was widely quoted in the media and then picked up, circulated, discussed and analyzed in policy, academic and other public circles. Books appeared bearing the metaphor on their covers. Newspaper journalists started using it in their editorials, college students in their term papers, Washington politicians in their speeches. Countless visualizations of the addiction trope appeared. These pictures often show Uncle Sam, the quintessential representation of the American body politic, as a worn-out drug addict. Figure 2.1 is a case in point. In this cartoon, published in 2010, President Barack Obama attempts to convince Uncle Sam to kick his oil habit in favor of alternative sources of energy—something, the cartoon suggests, that is easier said than done.

Just like the lifeblood trope discussed in the previous chapter, the notion of an American oil addiction is a body-politic metaphor. As such, it represents another fine example of how the hydrocarbon society has imagined oil in socio-medical terms. And just like the lifeblood trope, the addiction metaphor too contains deep-seated American assumptions about the country's relationship with its central source of energy.

The lifeblood and addiction tropes are central to understanding political definitions of dependency in the United States because they reflect two different approaches to the subject. That is to say, they symbolize in different ways how the hydrocarbon society has dealt with the inescapable fact that its existence is utterly dependent on the continued supply of oil. Oil as lifeblood describes petroleum as a natural and indispensable part of the body politic. Oil is healthy. It represents the nation's essence or, in the words of ad-man Bruce Barton, its "juice from the fountain of eternal youth".[4] Oil invigorates America, energizes and empowers it, defines its youthful strength. Oil dependence, it is implied, is a benign condition. Sure, Uncle Sam needs his lifeblood, but there is nothing inherently wrong with that.

The addiction metaphor, by contrast, mounts a full-frontal assault on this conceptualization of dependency. To begin with, oil is no longer imagined as being inside the body politic. Images using the idea of an American oil addiction usually depict oil as being injected into Uncle Sam's body from the outside. Sometimes Uncle Sam uses a needle to

Figure 2.1 Sam, you're addicted to oil, 2010. The notion of an American oil addiction dates back to the 1970s, but as a dominant way of imagining the nation's relationship with oil it became important only in the twenty-first century. Since then, it has done much to reshape the way in which the problem of dependency has been debated.[5]

pump oil into his veins like heroin, or, as in Figure 2.1, he swallows it like an ever-thirsty alcoholic. In either case, oil is no longer a natural, indispensable part of the body but something that has to be internalized. More than that, oil is something that does not actually belong inside a healthy body. Oil is artificial, unnatural and unhealthy—something dangerous. Oil-as-lifeblood equates petroleum with the vital force of the United States; oil-as-drug is the exact opposite of vitality: it is a poison that slowly consumes and ultimately destroys the body politic. Dependency, from this point of view, becomes an unbearable burden, and, as George W. Bush underlined in his speech, an urgent political problem that requires attention, deliberation and action by policy makers in Washington.

The assumptions reflected in the lifeblood and addiction metaphors have entered into political definitions of the problem of dependency as either foreign oil dependence or general oil dependence, but as the different histories of the two metaphors suggest, these definitions have changed significantly over time. The fact that both are currently widely used suggests, moreover, that the American discourse of dependency today is heavy with ambiguity, contradictions and uncertainty. As the following pages will show, in the hydrocarbon society prior to the 1970s dependency was not widely defined as inherently problematic. After the watershed of the 1973 oil shock, this changed abruptly and dramatically. Dependency was now seen as a major political problem, but—and this is the key point—primarily in the form of foreign oil dependence. In line with the representation of oil as lifeblood, foreign oil dependence was all about ensuring the steady supply of that vital material without which the hydrocarbon society could not exist. The addiction metaphor remained at the margins of the political debate in the 1970s, and became important only in the early 2000s. Its emergence echoes a gradual, ongoing process of reimagining oil in the United States. As environmental concerns and the threat of resource scarcity have become more pressing, *all* forms of oil dependence have increasingly been problematized in the United States. This, in essence, is what makes George W. Bush's oil addiction statement so interesting. The President had challenged the dominant definition of dependency before the entire nation. Both readings of dependency, foreign and general, co-exist in his statement in an ambiguous, uneasy relationship. Just as in Bush's speech, oil discourse in the United States today is shaped by two contradictory notions of dependency, as well as fundamental uncertainty about the primary meaning of this political problem.

This chapter offers a brief history of how the meaning of dependency has been constructed out of the cultural raw materials presented in the first chapter. The best way of writing such a history of dependency as a political problem is to trace its genealogy. To do this is to examine the formation of a political problem over time. As a mode of historiography, genealogies are not primarily concerned with studying the events of the past but with writing, as Michel Foucault put it, "a history of the present".[6] Jens Bartelson succinctly explains this concept: "A genealogy has not as its task to tell what actually happened in the past, but to describe how the present became logically possible."[7] From this perspective, the American problem of dependency should be understood as a series of discourses, each with its own context, logic, content and language. They are studied here with the ultimate aim of illuminating how the discourse of dependency operates in the United States in the present—with all its unquestioned associations and implications, its ambiguities and contradictions.

From the Beginning to the 1970s

It is perhaps impossible to determine with any degree of certainty when the United States first became dependent on oil. In a straightforward material sense, the emergence of this dependence is closely linked to the spread of oil and oil-based fuels in US energy markets and the increasing availability of internal combustion engines, especially those in cars, in the first third of the twentieth century. By the 1920s, certainly, oil's central role in the US economy was there for anyone to see. Two British scholars visiting the United States at the time had this to say about their trip: "Travel but little in the country and you will gain the impression that the modernism of the United States flowed from its oil wells [. . .] Does not the American partly live in oil?"[8]

Naturally, American observers were not oblivious to the increasing importance of petroleum products for the country's consumers. The following lines from a 1935 article in the *Saturday Evening Post*, then the country's leading weekly magazine, exhibit an acute awareness of this issue:

> There is no doubt about our absolute and complete dependence upon oil. We have passed from the stone age, to bronze, to iron, to the industrial age, and now to an age of oil. Without oil, American civilization as we know it could not exist. No wheel could turn, no machine could function. No part of our motorized transport system could budge. Our entire system of distribution, our very habits as individuals, float upon oil.[9]

The author of the article was Harold Ickes, Roosevelt's secretary of the interior between 1933 and 1946 and a central figure in national energy policy making of the time. In his article, Ickes seized on the time-honored notion that energy and the progress of human civilization were intimately connected. Oil dependence, he seemed to suggest, was an inevitable part of this new age. But, and this is the main point here, this in itself did not make dependency a problem. The problem, according to Ickes, was waste.

In order to understand this, it is necessary to grasp the historical juncture at which Ickes published his article. Since the early 1900s, the issue of conservation of oil resources had been debated in industry and policy circles. In the 1920s and early 1930s, questions of resource conservation, price stabilization and production control acquired new urgency in the face of rampant overproduction of oil. The discovery of new fields—chief among them the gigantic East Texas field—led to what Yergin calls an "orgy of drilling".[10] In 1931, domestic oil production stood at 2.5 million barrels per day, compared to market demand for only about 2 million barrels. Prices dropped 50 percent below the 1929 mark, threatening not only the oil industry but the entire American economy.[11] In this context, Ickes' concern was not dependency but the large-scale squandering of a natural resource recognized as important. "I challenge any other present-day industry in the United States", he wrote, "to show greater waste, inefficiency and mismanagement than seem to be inherent in the oil industry." During the time it took the reader to finish his article, the Secretary declared, enough oil was being wasted to keep at least 10,000 American families warm for an entire winter.[12]

Ickes' article, and the fledgling conservation debate in general, provide telling insights into American readings of dependency in the early hydrocarbon society. Proponents of conservationism could agree that petroleum was essential to modern life in the United States, but this was not widely seen as a political problem.[13] The insight that oil was ultimately a finite resource and that, one day, the country's reserves would be depleted did not lead them to call for a new energy foundation for the United States. As far as dependency as a political problem was concerned, therefore, there was mostly silence. If oil dependence was represented in explicit terms, this was done in line with the assumptions contained in the lifeblood metaphor. Oil appeared as an invigorating substance that enabled material progress and a better life in the United States. Dependency on petroleum was just a plain, simple fact of life.

The real oil issue at the time, as Timothy Mitchell has suggested, was the problem of abundance. Policy makers and industry experts were looking for ways of governing the production of a commodity that was available in such unimaginable quantities that abundance itself became the central concern. In this context, dependency was not only not seen as a problem but increasing oil reliance could even serve as a solution, as a "method of preventing energy abundance".[14] Ultimately, policy makers decided to deal with abundance by pro-rationing production—that is, by limiting production levels to expected market demand. Between the 1930s and the 1970s, the United States operated such a system of pro-rationing oil, with the so-called Texas Railroad Commission at its center. The goal was to stabilize excessive and volatile production levels according to free-market demand.[15]

In the early hydrocarbon society, dependency was therefore not widely articulated by policy makers as inherently problematic. Even more fundamentally, public debates in the United States were generally shaped by silence about big energy questions. To the extent that oil became a matter of political debate, discussions usually remained confined to specialized political, academic or industrial circles. There were a few instances in which oil moved into the public spotlight, including the dissolution of the Standard Oil Trust in 1911, the Teapot Dome scandal in 1923 and the gasoline-rationing policies implemented during World War II. By and large, however, questions about oil monopolies, the regulation of competition in the industry, the extent of domestic oil reserves, control of oil leasing rights, industrial waste, pro-rationing, conservation or oil imports were not the kinds of issues that Americans cared about. As Richard Vietor observes, "the larger consuming public did not get involved as long as the surplus of oil kept prices constant".[16] Abundance was assumed to be the natural order of things, and as long as this expectation remained undisturbed oil remained a non-issue. If the public statements of US presidents can be considered a measure of whether a topic is worth debating by society at large, oil certainly did not qualify. Despite its overwhelming significance for life in the hydrocarbon society, State of the Union addresses and other major presidential speeches before the 1970s hardly ever contain even cursory references to oil. Only once did a sitting president make a passing remark about the importance of oil imports to Western Europe: Eisenhower in his 1957 State of the Union address. American dependence on oil was not mentioned, let alone described as a political problem.[17]

The Energy Crises of the 1970s

The 1970s are remembered today as a decade shaped by a sequence of energy crises. The two oil shocks, of 1973 and 1979, are central events in US energy historiography, but significant energy-related problems had begun to rock the American energy system much earlier. The notion of an "energy crisis" had already entered the American oil discourse by 1970.[18] By then, energy supply difficulties had become a continuous problem leading to rising energy prices and brownouts in parts of the country. The reason was simple: against a background of steadily rising oil demand, domestic oil production had peaked and then started to decline in 1970. The spare production capacity that had constituted the country's potential as a supplier of last resort was depleted soon afterwards. In April 1972, the pro-rationing authorities of Texas and Louisiana, which had controlled domestic production by determining quotas for the US oil industry since the 1930s, allowed drilling at 100 percent of market demand. This decision, in effect, made the practice of pro-rationing obsolete. In order to meet America's ever-growing energy hunger, President Nixon was forced to abolish government restrictions on oil imports in April 1973. These restrictions, known as the Mandatory Oil Import Program, had been established after much controversy under Eisenhower in 1959 and restricted oil imports in order to shield the American economy and the independent oil producers from an unregulated influx of cheap and abundant foreign oil.[19] According to Richard Vietor, the program had been "the single most important energy policy in the postwar era".[20] Now it was gone, and unprecedented quantities of foreign oil flowed into the United States to provide the energy that consumers and industries demanded and US domestic producers were increasingly unable to deliver. Net imports exploded from 3.1 million barrels per day in 1970 to more than 6 million barrels per day in 1973, about 35 percent of total US oil consumption.[21]

The end of pro-rationing and import quotas meant that two central pillars of the US system of regulating abundance had disappeared. Surplus management was no longer needed in a world that seemed to be increasingly defined by energy scarcity and shortage. In a secret study for the State Department in 1971, the department's top energy official, James Akins, observed that the world was passing through the "last gasp in the buyers' market".[22] The emergence of a sellers' market in oil brought with it important adjustments within the international energy regime. One was that price-setting power shifted from consuming countries toward the producers organized within OPEC. Another was that the United States became increasingly exposed to the cyclical ups and downs in the international oil market.

In this new environment, the two oil shocks of the 1970s impacted the United States with unprecedented force and directness. The first oil shock was a conjunction of several factors: firstly, the decision by OPEC's Persian Gulf members to raise the posted price of oil by 70 percent, to $5.12 a barrel, in an ongoing effort to gain control over the production and pricing of their oil resources; secondly, the Watergate Scandal in the United States; thirdly, the Yom Kippur War of October 1973; and, fourthly, the related decisions by nine Arab countries, all of them members of the Organization of Arab Petroleum Exporting Countries (OAPEC), to cut oil production and to impose an oil embargo against consuming countries seen as supporting Israel in the war.[23] A total of 4.5 million barrels of oil per day were removed from the market, about 13 percent of OPEC's pre-embargo production. Two earlier attempts by Middle Eastern producers to employ this "oil weapon", in 1956 and 1967, had been without significant effect largely due to the availability of US spare production capacity. By 1973, however, the United States had lost the ability to make up the shortfall. This, together with panic in the consuming countries and a heavy-handed US policy response, turned a minor shortage into a full-fledged crisis.

The second oil shock, in 1979, again brought together political events in the Middle East with a desire by the producers to achieve increased oil revenues. It too caused panic in the consuming countries. During the Iranian Revolution, Iran's oil production dropped heavily. Although there was no actual shortage of oil as other producers stepped up their production, fear of shortages and possible upheaval in other producing states caused panic and over-buying in the markets.[24]

In order to grasp the significance of these events for the hydrocarbon society, we need to enquire about the nature of an "oil shock". The term has its origins in the field of economics, where an economic shock is understood to be a sudden, unexpected and exogenous event that causes change in key macroeconomic variables. It is a conventional assumption that the combined effects of OPEC's price hike and the OAPEC embargo helped trigger an economic downturn in the United States.[25] In November 1973, the US economy moved into a recession that lasted for 16 months. In 1974, GDP contracted by 0.5 percent, followed by another 0.2 percent in the following year.[26] For the first time, the US economy was haunted by stagflation—a term denoting stagnating output coupled with high levels of inflation. According to Bruce

Beaubouef, "oil supply disruptions and price hikes, imposed on the US from without, had a very real impact upon the national economy". Inflation rates hit 10 percent at times, a third of which, Beaubouef argues, could be attributed to rising energy prices. High oil prices amounted to a tax on Western consumers, causing an unprecedented transfer of wealth from the industrialized, oil-consuming West to the underdeveloped oil-producing countries. Inflation ate into the disposable income of Western middle-class consumers, and real wages ceased to grow for the first time since 1945. From November 1973 to March 1975, purchasing power in the United States decreased by about 9 percent. Because petroleum was essential to many key industries in the United States—not just car makers but also the transportation and construction sector—the price hike affected productivity in many areas. About 500,000 US workers were laid off as a consequence.[27]

To characterize the events of the winter of 1973 and 1974 as an economic shock risks losing sight of an equally important point. The embargo also, and perhaps more importantly, constituted a shock to oil's *discursive* economy in the United States—that is, to the cultural foundations of the hydrocarbon society. It was a discursive shock as much as an economic one. In October 1973, the embargo hit a nation that was already in upheaval over the Watergate Scandal and the Vietnam disaster. When gasoline prices shot up by 40 percent, long "gas lines" appeared, and fuel was rationed and sometimes unavailable altogether. The everyday routines of American life—filling up the tank of one's car without wasting a thought about oil supply—suddenly became something that people needed to worry about. Many responded by buying as much gasoline as possible in advance, thereby creating even greater shortages. Some Americans could no longer afford the gas they needed to get to work. Rising heating-oil prices meant that some could no longer afford to heat their homes. Uncertainty was pervasive and conspiracy theories mushroomed. Was the crisis real? Or was it a fabrication by the oil companies or even the government in Washington? "Things meshed in hazy, mysterious ways", Yergin writes, "and this impression left deep and abiding suspicions that fed conspiracy theories and obstructed more rational responses to the energy problems at hand."[28] Fist fights and even shootings at gas stations were reported as American motorists were gripped by anger and panic. The question of energy supply had turned into a full-fledged social crisis which, in the words of Bruce

Beaubouef, "pitted region against region, producers against consumers, economic class against economic class. American society was being rent asunder."[29]

The energy crisis was a shock to the cultural foundations of the hydrocarbon society because it rendered precarious the entire corpus of unquestioned, tacit assumptions about the role of oil in American society. More than that, it called into question the very foundations of economic and social life in the United States, and forced Americans to reconsider and evaluate that which they had long ago accepted as given. "The crisis threatened", as Karen Merrill writes in her history of the oil shock, "the first principle of America's astounding economic growth: access to cheap oil and gasoline. In other words, the oil crisis cut to the core of American assumptions."[30] The unfolding events called into question the very notion of abundance which had so far formed the centerpiece of oil culture in the United States. The energy shortage was, Rutledge notes, a "shock to America's image of itself as an oil rich nation."[31] In much the same vein, Yergin argues that

> the shortfall struck at fundamental beliefs in the endless abundance of resources, convictions so deeply rooted in the American character and experience that a large part of the public did not even know, up until October 1973, that the United States imported any oil at all.[32]

There had been times before when energy appeared scarce, but shortage had always turned into renewed abundance soon after. This time it seemed different. This time, James Akins wrote, the wolf was here.[33] "The days of cheap, abundant energy were slipping away, and with them a sense that America the Abundant was slipping away too."[34]

In the cultural history of the hydrocarbon society, then, the energy crises of the 1970s must be considered as a watershed, a key moment of crisis. The first oil shock, in particular, demonstrated that the narratives and assumptions about oil that had formed the key building blocks of US energy culture were not immune to challenges, thus illustrating the theoretical insight that the construction of meaning is inherently unstable. Meaning, as Stuart Hall argues, "has constantly to be renewed". Under certain circumstances, it can "disappear or be overthrown, leading to the old linkages being dissolved and new connections—

re-articulations—being forged."[35] Such change can occur gradually and largely unnoticed, or it can happen abruptly and violently. In the latter case, we can speak of a discursive shock. Such an event opens up the possibility for new assumptions to challenge, marginalize or undermine taken-for-granted descriptions, self-evident truths and common wisdoms. At the extreme, perhaps, such discursive shocks may give rise to what Alexander Wendt calls "moments of reflexivity", fleeting moments in which an entire society is able to grasp the nature of the world as a social construction, see through its own reifications, remember its authorship of the world and embrace a different perspective.[36] The historical conjuncture of the energy crisis provided precisely this kind of discursive crisis, with consequences for the way in which Americans have imagined themselves in relation to energy that continue to shape the hydrocarbon society to this day.

What changed? The most obvious consequence of the energy crisis was that it laid bare the cultural foundations of energy use in the United States. Assumptions which had so far remained implicit—contained, for example, in the lifeblood metaphor—were now articulated and questioned consciously and openly. Moreover, for the first time such debates had a truly public character. As suggested above, big energy questions had not usually been a major concern outside specialized industry and administrative circles. This changed in the early 1970s, when the notion of an energy crisis became central to the political vocabulary of the news media and the general public. In June 1971, President Nixon sent a special message on energy to Congress, the first ever major presidential statement on energy.[37] As a political issue, oil had not enjoyed this level of urgency and visibility since the nationwide rationing of gasoline in the 1940s. And then, in 1974, more than a century after it was first struck on American soil, oil finally gushed up to the supreme heights of American politics when President Nixon made it the main subject of his State of the Union address. The newness of oil as a major political issue was not lost on the President. "In all of the 186 State of the Union messages delivered from this place", Nixon declared, "in our history this is the first in which the one priority, the first priority, is energy."[38] For the first time, America's relationship with oil had become an object of truly public political consideration, an issue which required deliberation and interpretation before the entire nation. Even the President interpreted this sudden change in terms of a shock to the core assumptions about the abundance and limitlessness of energy in the United States. As he explained:

> For most of our history, a plentiful supply of energy is something the American people have taken very much for granted. In the past twenty years alone, we have been able to double our consumption of energy without exhausting the supply. But the assumption that sufficient energy will always be readily available has been brought sharply into question within the last year.[39]

The first consequence of the oil shock, then, was to render impossible the continuation of silence about the limited energy foundations on which the American way of life was based. With the discourse of abundance in crisis, the notion of dependency suddenly became a central concern. After at least four decades of fundamental American reliance on oil, dependency finally emerged as an object of politics. In the wake of the energy crisis, dependency was almost universally articulated as the most critical energy-related problem facing the United States. However, and this is the crux of the matter, two distinct readings of dependency emerged during the 1970s. According to the first view, the problem was American oil reliance at large. Dependency was associated with excessive patterns of domestic oil consumption and with wasteful and inefficient use of natural resources that were valuable and scarce. This definition of the problem was closely connected to the emergence of "the environment" as a new area of political concern.

In the wake of the oil embargo against the United States in October 1973, however, the problem of dependency was increasingly articulated in different terms. Since the oil shock had been inflicted on the country from outside, policy makers offered a reading of the situation that emphasized America's dependence on *foreign* oil. In government circles and well beyond, foreign oil dependence, not reliance on oil as such, was accepted as the central energy-related political problem and an issue of national security that needed to be addressed urgently. This definition of dependency sparked into life a new discourse about the aspiration of "energy independence", which was defined as a solution to the dangers of foreign oil. In the eyes of the American public and Washington policy makers, the second oil shock of 1979 only seemed to confirm this reading of the problem.

Dependency as General Oil Dependence

Uncertainty about the origins, reality and implications of the energy crisis undermined what had been taken for granted about energy in American society and opened up a void that needed to be filled with meaning. "What is striking in hindsight", writes Karen Merrill, "is the depth and complexity of the self-scrutiny that many Americans undertook to grasp what had happened."[40] The entire range of discourses about abundance, growth, progress and freedom on which the hydrocarbon society had been constructed came under intense scrutiny. Deep introspection filled the pages of influential newspapers such as the *New York Times*, which asked its readers: is unbridled growth indispensable to the good life?[41] The paper was hardly alone in raising such weighty questions. Was the notion of limitless natural resource abundance and endless economic growth really a solid foundation for modern life? Was bigger really better? Of the many critical enquiries into the basic assumptions underpinning the American way of life, few were as influential as the Club of Rome's study *The Limits to Growth*. The authors had used computer models to predict the consequences of current trends in such areas as resource depletion, population growth, industrialization and environmental pollution over the course of the twenty-first century. Their central argument—that the prevailing system was bound to collapse if nothing changed—represented a full-frontal assault on the notion that limitless growth was possible or even desirable.[42] This new logic of limits challenged a more optimistic view of material progress that had been an important feature of the hydrocarbon society after 1945. Paul Ehrlich's bestseller *The Population Bomb* warned about the dramatic consequences of unchecked population growth in a finite ecology, while E. F. Schumacher's *Small Is Beautiful* argued, among other things, that natural resources should no longer be treated as if they were endless.[43]

The role that energy consumption played in driving growth and material progress did not escape this kind of skeptical scrutiny. If increasing energy consumption had previously been seen as essential to sustain the American way of life and the promise of a better future, some now began to question the validity of this connection and the sustainability of current energy-consumption patterns. "Why must energy and welfare march forever in lockstep?", asked environmentalist Amory Lovins in a widely read paper, and called for a "softer energy path" for the country based on decentralized non-fossil energy supply.[44]

Criticism of the key tenets of American society was nothing new, of course. The country's consumer culture and the assumption of limitlessness had all come under attack before.[45] Now, however, enquiries into the foundational values and assumptions of American life coincided with the emergence of a new field of political concern: the environment. The adverse environmental effects of producing and consuming fossil fuels were one of the issues around which the fledgling environmental movement began to crystallize in the 1960s. A number of air pollution crises prompted utility companies to shift from coal to oil, but petroleum too was being linked with threats to the natural environment. The Santa Barbara oil spill became a potent symbol for these dangers. On 28 January 1969, an oil well operated by Union Oil on a federal lease off the California coastline blew out, releasing some 230,000 gallons of crude oil into the sea. The crisis sparked a range of political investigations, hearings, studies and administrative measures. It helped make the environment an object of politics, and one that rivaled the status of the economy as a framework for assessing the meaning of energy issues.[46] The spill moved the newly established Nixon Administration toward adopting the National Environmental Policy Act (NEPA) with its stated aim of protecting and enhancing the natural environment. Through events like the Santa Barbara incident, oil became increasingly "dirty". By the end of the 1960s, Melosi observes, "oil was no longer identified only with material progress, mobility, industrial growth, and national security; it was also linked to health risks, environmental despoliation, and overconsumption".[47]

In this context, new descriptions of American reliance on oil appeared which defined dependency as a serious political problem. A number of environmentally conscious academics, activists, intellectuals and others began to characterize the extent to which everyday life in the United States relied on oil consumption as excessive, problematic, shameful and unsustainable. In other words, the American way of life was seen as the primary cause of the country's energy predicament. The following account of the energy crisis from a 1973 *Foreign Affairs* article illustrates this reading of the crisis:

> The United States is facing a national energy emergency. It arises from our extravagant and wasteful use of energy and from a shift in the sources of fuels. Per capita consumption is three times that

> of Western Europe, and we may ask ourselves whether our greater use enriches the quality of life by any such margin. Our cars are twice as heavy and use twice as much fuel as European cars which run about the same mileage each year [. . .] We keep our houses and buildings too hot and use large amounts of fuel in air-conditioning everything. We have not given a thought to fuel conservation and efficiency since the days of rationing in World War II [. . .] These are some of the reasons why with six percent of the world's population the United States uses 33 percent of the world's energy.[48]

Such definitions of oil reliance as inherently problematic were also implied in addiction metaphors, which now began to appear. Oil was described as "that vanishing, magic substance to which the American people have become slavishly and hopelessly addicted".[49] To speak about oil in terms of an addiction was another way of saying that the energy crisis had its origins in the profligate squandering of energy by Americans. Natasha Zaretsky writes that "the image of the nation as junkie conveyed a deep sense of shame about the newly compromised position of the United States and, more specifically, about the out-of-control nature of American oil consumption". Dependency, she argues, was seen as posing an essential threat to the lifestyle and moral authority of the white middle-class family.[50]

Before the crisis, few people saw the need for the nation to stop using oil. Now, however, analysts, scholars and policy makers began to ponder the possibility and necessity of a third American energy transition: away from oil and toward more sustainable, renewable energy sources. Like the notion of an oil addiction, the debate about another energy transition contained within it an implicit but striking critique of the idea that oil constituted America's essential, unproblematic lifeblood. Transition proposals reflected an emerging political definition of dependency which emphasized the dangers inherent in all forms of fossil-fuel consumption.[51]

The discursive shock of the energy crisis opened up space for these new assumptions to seep into official representations of the events. The notion of an energy transition, for example, occasionally appeared in the statements of Gerald Ford. In 1974, the President outlined his vision for a time after the hydrocarbon society: "The power of the atom, the heat of the sun and the steam stored deep in the Earth, the force of the winds and water must be main sources of energy for our grandchildren, and we can do it."[52] Under his successor, Jimmy Carter, these ideas entered the inner sanctum of public policy. Today, Carter is remembered, among other things, for labeling the energy crisis "the moral equivalent of war" and for his famous "Malaise" speech.[53] Equally important was that Carter described the energy problems of the United States in terms of excessive demand, over-consumption and waste. In doing so, the President espoused many of the cultural criticisms put forward by intellectuals such as Robert Bellah, Daniel Bell or Christopher Lasch, which he elevated to the supreme heights of US presidential discourse.[54] Like his predecessors, Carter depicted dependency as a problem of ensuring the supply of imported oil from unreliable sources, but more than Nixon and Ford he went beyond that to argue that wasteful practices of oil consumption were at the heart of the problem as well. As the President frequently stated, "We simply use too much oil, we waste too much oil, we import too much oil."[55] Moreover, Carter was firmly convinced—"clearly obsessed" according to Francisco Parra—that the world was about to run out of oil.[56] The necessity of another energy transition was therefore an urgent political reality in Carter's view. In his 1979 energy address, the President set out before the nation a broad energy historiography in which he linked American exceptionalism and the nation's unmatched living standards to its two previous energy transitions. Carter concluded, "Because we are now running out of gas and oil, we must prepare quickly for a third change—to strict conservation and to the renewed use of coal and to permanent renewable energy sources like solar power."[57] In the President's narrative about the crisis, there was no trace left of the confident discourse of abundance that had previously characterized American oil culture. Emerging notions such as scarcity, limits, waste, pollution, over-consumption and addiction were challenging older views of American reliance on oil as inherently benign.

Dependency as Foreign Oil Dependence

Today, the energy crisis is remembered in the United States primarily as a disruption of the country's foreign oil supply. More specifically, the complex conjunction of forces that constituted the crisis is often reduced to one primary cause: the oil embargo against the United States and other countries in response to American foreign policy in the Yom Kippur War.

The war had started on 6 October 1973, when Syrian and Egyptian forces attacked Israel in an attempt to regain control of territory occupied by Israeli forces since 1967. Both Cold War superpowers quickly became involved in the conflict, seeking to contain the war and, at the same time, to avoid the complete military defeat of their respective regional protégés. Despite Saudi Arabian warnings that US intervention in support of Israel might lead to an oil boycott, Washington began to supply the Israeli forces with arms and ammunition. On 17 October 1973, the oil ministers of ten Arab countries announced that their oil production would be lowered by 5 percent each month until all Israeli forces had withdrawn from the territories occupied in 1967. Three days later, in reaction to continued US support for Israel, Saudi Arabia announced that all further oil shipments to the United States would be suspended. Over the following days, all Arab producers took similar decisions against the United States and other countries seen as pro-Israeli. The embargo was in effect and continued even after a ceasefire was agreed between the warring parties on 25 October. On several occasions during the winter of 1973–4, Arab oil ministers met to amend the terms of production cuts and shipment bans, and it was not until 18 March 1974 that the embargo was finally ended.[58]

In the United States, the oil embargo triggered a redefinition of the meaning of the energy crisis. In January 1974, Richard Nixon sent a special message to Congress about the events. The President claimed that the problems experienced by Americans had originated outside the United States. Nixon admitted that "shortages were long in appearing" but insisted that "the energy crisis itself came suddenly, borne by a tragic war in the Middle East". Nixon then offered a narrative that was radically different from the reading of the crisis contained in addiction metaphors and energy transition proposals. He declared:

> We learned, at a stage short of the truly critical, that we had allowed ourselves to become overly dependent upon foreign supplies of a vital good. We saw that the acts of foreign rulers, even far short of military action, could plunge us into an authentic crisis. The Arab oil embargo will temporarily close some gasoline stations, but it has opened our eyes to the short-sighted policy we had been pursuing.[59]

Nixon's explanation of the country's energy problems as a "foreign-made crisis" formed the nucleus of what was to become the dominant interpretation of the problem of dependency. This definition accepted the importance of oil to American society ("a vital good") and even recognized that increasing American demand entailed practical problems of ensuring supply. But that was not the main issue. Instead, the central problem—the problem of dependency—was to ensure supply of foreign oil, most notably from the unreliable producers in the Middle East. In this view, dependency meant foreign oil dependence. The energy crisis was a crisis of foreign oil supply. There was no challenge in Nixon's words to the notion that oil was America's lifeblood, no trace of an American addiction. The President described the country's energy requirements in terms of innocent American "energy needs" or "essential civilian demand", which needed to be protected against economic blackmail by the producers or accidental disruptions caused by wars in the Middle East. On another occasion, Nixon offered a concise summary of this reading of the crisis. "We must never again be caught", he declared, "in a foreign-made crisis where the United States is dependent on any other country, friendly or unfriendly, for the energy we need to produce our jobs, to heat our homes, to furnish our transportation for wherever we want to go."[60]

This definition of dependency became the core of official representations of the energy crisis, and thus provided the basis for US policy reactions to it. What is more, an official emphasis on the problems associated with foreign oil dependence survived the spectacular implosion of the Nixon Administration in the summer of 1974. Speaking three years after the embargo, President Ford defined dependency in exactly the same terms: "The principal energy problem now facing the United States is our excessive and growing dependence on imported oil from a relatively few foreign nations that own the majority of world oil reserves and have the ability to control world oil prices and production."[61] Jimmy Carter's view of energy, as explained above, was more complex. He depicted US overconsumption and the dangers of importing oil in roughly similar terms. "The fact is", Carter explained, "that by leaps and bounds the American people are importing and using too much oil. This has been the primary cause for our concern."[62] Foreign oil dependence was a significant part, but just one part, of "the greatest challenge that our country will face in our lifetime", requiring a national effort the President described as "the moral equivalent of war".[63] What these official declarations about the problem of dependency had in common was that they causally linked the American practice of importing oil to a range of very serious risks, threats, dangers and potential catastrophes. As the following chapters will show, the dangers of foreign oil included the risk of economic disruption, the loss of US political independence and sovereignty, a threat to the status of the country as the world's leading superpower and a collapse of international order itself.

To be sure, the argument here is *not* that foreign oil dependence emerged as an entirely new political issue in 1973. Concerns about foreign oil were formulated in internal US government documents as early as 1919, and continued to be discussed throughout the following decades.[64] In 1970, President Nixon commissioned a cabinet-level task force with assessing the national security implications of oil import policies. The final report produced by this task force further demonstrates the purchase of security arguments within the administration prior to the embargo.[65] The significance of the discursive shock of 1973 lies not in the emergence of entirely new definitions of dependency. Its principal effect was to transform an already existing concern about the dangers of foreign oil into the central energy-related object of political anxiety in the United States. Worries about foreign oil had previously been found mostly among experts; now politicians, the media and the public moved the problem of dependency into the heart of American political discourse. Speaking about oil imports in terms of threats to national security achieved the status of a common-sense, self-evident description of the energy situation in the country far beyond the boundaries of administrative and industry debates. Capturing this sentiment, the *Washington Post* described reliance on foreign oil as "the biggest, most painful single problem ever met with by the US in peacetime".[66] The American public, by and large, shared this view. In a 1977 nationwide poll, for example, 83 percent of those surveyed answered "very important" when asked how important it was to "try to reduce US dependence on foreign oil".[67] Perhaps even more telling than these figures were some of the survey questions: "How confident are you that we can significantly reduce US oil imports during the next few years?"[68] "Which of these energy sources do you think are realistically possible to use for replacing foreign oil during the next five years?"[69] The claim that foreign oil dependence was a dangerous condition facing the country was already prefigured in the way the questions were phrased.

Both those who formulated and those who answered them could agree on this basic description of the situation.

The significance of the discursive shock of 1973–4 thus lies in shaping a definition of dependency which was to become the most viable frame for discussing energy in the United States. A reading of the energy crisis that described the problem as one of foreign supply rather than domestic demand was in line with the central cultural tenets of the hydrocarbon society. The American discourse of abundance contained a narrative according to which shortages of energy were not natural but represented artificial limitations of supply.[70] In 1973, this narrative lent itself to the view that the energy crisis represented an "aberration that came from abroad".[71] It is perhaps not surprising, therefore, that defining dependency as a problem of ensuring foreign oil supply quickly crowded out the alternative of seeing dependency as a problem of domestic over-consumption.

The Promise of Energy Independence

During the energy crisis, the US Government outlined the goal of energy independence as a policy response to the problem of dependency. The promise of energy independence—to be free again from the dangers of foreign oil—demonstrates the extent to which the energy crisis and foreign oil dependence had become equated. All three US administrations of the 1970s declared their intention to pursue such a policy. It had already appeared in Nixon's statements as "self-sufficiency" early in 1973, but a forceful public policy announcement was only made in November as a direct response to the oil embargo. In a speech on how to deal with the energy crisis, the President announced a great new national project:

> Let us unite in committing the resources of this Nation to a major new endeavor, an endeavor that in this Bicentennial Era we can appropriately call "Project Independence". Let us set as our national goal, in the spirit of Apollo, with the determination of the Manhattan Project, that by the end of this decade we will have developed the potential to meet our own energy needs without depending on any foreign energy sources. Let us pledge that by 1980, under Project Independence, we shall be able to meet America's energy needs from America's own energy resources.[72]

Literature on the energy crisis does not usually take the energy independence policies of the 1970s very seriously. In Yergin's account, for instance, Nixon's Project Independence appears as an attempt to divert attention away from the brewing storm over Watergate.[73] Many also point out that the project was inherently unfeasible. According to Parra, "the idea was wildly unrealistic, and everybody in energy circles knew it".[74] Even the official *Project Independence Report*, published by the Federal Energy Administration in November 1974, concluded that foreign oil imports could be limited but not entirely avoided without severe macroeconomic costs.[75] Irrespective of such damning criticism, Gerald Ford adopted the policy with some modifications.[76] Both Jimmy Carter and Ronald Reagan, while avoiding formal energy independence deadlines, continued to formulate policies aiming to achieve energy independence, which they understood as "energy freedom" or "energy non-dependence". By this they meant continued imports but without the economic and political vulnerabilities concomitant with dependence.[77]

Like descriptions of the problem of US foreign oil dependence, promises of energy independence were saturated with references to the key building blocks of American national identity. This is already apparent in Nixon's announcement quoted above, but became even more explicit in another speech which the President gave three weeks later:

> Let me conclude by restating our overall objective. It can be summed up in one word that best characterizes this Nation and its essential nature. That word is "independence". From its beginning 200 years ago, throughout its history, America has made great sacrifices of blood and also of treasure to achieve and maintain its independence. In the last third of this century, our independence will depend on maintaining and achieving self-sufficiency in energy.[78]

The quest for energy independence does not appear as a new goal in Nixon's narrative but as a return to the normal state of national independence. Foreign oil dependence, the President explained, constituted an aberration from the American state of nature. Project Independence was a struggle to reconquer this paradise lost of American independence.

Project Independence might have been an unrealistic policy but it should still be taken seriously because it demonstrates the power inherent in the emerging definition of dependency as foreign oil dependence. The promise of energy independence invariably refers to making the United States secure from the dangers of foreign oil. As a policy strategy, it is thus fundamentally different from the notion of an energy transition away from oil that, while also freeing the United States from the need to import oil, is based on a much broader reading of dependency in which all forms of oil consumption are seen as inherently problematic. As a course of action, in other words, the promise of energy independence was inextricably linked to the prevailing definition of foreign oil dependence as the nation's key energy problem.

This, then, is the central finding emerging from an examination of the 1973 energy crisis as a discursive shock. The crisis opened up a brief moment of reflexivity in which Americans pondered and questioned the energy foundations of the hydrocarbon society. Different readings of the crisis emerged, as the co-existence of the addiction and lifeblood metaphors demonstrate. In the wake of the embargo, however, the Nixon administration offered a reading of the events which emphasized the foreign origin of the crisis and depicted the country's dependence on oil imports as the key cause of the troubles the country was experiencing. The aspiration of energy independence became the logical solution to this problem. This reading not only came to dominate American debates about energy, it also persisted long after the direct effects of the embargo had faded. Its validity as a description of the world seemed to be confirmed to Americans when the second oil shock hit the nation in 1979–80 in the wake of the Iranian Revolution.

Since the 1970s, in effect, the twin notions of foreign oil dependence and energy independence have constituted one central discursive foundation for American debates about energy. They became what Stefan Halper and Jonathan Clarke call "Big Ideas". US foreign policy, Halper and Clarke observe, has historically been formulated on the basis of a range of themes that linked policy issues to the key tenets of American identity, most importantly the notion of American exceptionalism. The authors explain:

> The Big Idea is a rhetorical device that rests on shared assumptions reflecting American Exceptionalism and functions as a direction-

> setting form of shorthand, an easily grasped metaphor or signpost for an underlying notion that is difficult or tiresome to spell out or, indeed, may not be well understood.[79]

It was during the energy crisis of 1973–4 that saying "foreign oil dependence" became shorthand for evoking a range of dangers and threats to the United States, while saying "energy independence" became shorthand for the promise of making the country invulnerable to these dangers.

Dependency After the 1970s

The energy crises of the 1970s challenged cultural assumptions underpinning the hydrocarbon society but did not destroy its material foundation. Oil continued to shape life in the United States. The share of petroleum in total US energy consumption gradually shrank from 44 percent in 1980 to about 36 percent in 2013, but oil remains the most important primary energy source for the United States today. The United States was, and continues to be, fundamentally dependent on oil. More to the point, despite many solemn promises of energy independence, the United States never ceased to import tremendous quantities of oil. In fact, after a brief period of decline, which lasted until about 1985, imports began to rise steadily. The United States began the 1980s by importing 37 percent of its total petroleum needs. This figure had risen to 42 percent by 1990 and then to 53 percent by the year 2000. For 2005, US Energy Information Administration (EIA) statistics record a peak of US foreign oil dependence at about 60 percent of total demand, the equivalent of 12.5 million barrels of oil per day. Since then, imports have been in what looks like secular decline. Still, in 2013 domestic oil production could only satisfy 77 percent of US oil consumption. The remaining 6.2 million barrels per day came from abroad.[80]

Redefining Dependency as an Addiction

Definitions of the problem of dependency emphasizing unsustainable and wasteful patterns of domestic energy consumption were formulated in many quarters in the wake of the discursive shock of the energy crisis. Unlike the official narrative of the energy crisis, which centered on foreign oil dependence, this reading of dependency was actively opposed by some elements of US society, most notably the energy industry. The big oil companies consistently worked to dispel the notion that Americans were consuming more than their fair share of energy. Mobil's "Op-Ed" series of editorial-style adverts in the *New York Times* and other big newspapers, for example, repeatedly lashed out at what the company saw as attempts to create "a national guilt complex" around energy consumption. As one advert argued, "Nobody in this country has to beat himself over the head just because he's adequately fed and clothed."[81] The company agreed that it was worthwhile to encourage people to save energy but that policies aiming to force Americans to do so were fundamentally inconsistent with the birthright of the American people to make free choices. Another advert explained:

> Frankly, we don't feel the private homeowner should be made to feel guilty for his style of accommodation, especially since a home of one's own is so integral a part of the American dream. We agree that conservation can contribute to an energy program. But conservation should mean eliminating waste. It should certainly not become a moral issue, and federal administrations should not become the arbiters of public morality [. . .] An energy policy that attempts to legislate morality goes too far.[82]

What Mobil and the other companies advocated instead was a supply-side approach to solving the crisis. Based on the old industry claim that abundance was the natural result of free-market capitalism, the companies argued that there was more than enough oil and that all that stood in the way of cheap and abundant energy were artificial political and administrative obstacles.

With Ronald Reagan, elected president in 1980, belief in abundance also re-emerged in the White House. The President flatly denied that the crisis had had anything to do with domestic shortages. "The truth is", Reagan proclaimed on the campaign trail, "America has an abundance of energy."[83] The new president intended to solve the country's energy problems through supply-side measures, not through intervention in domestic demand. Americans regained faith in the energy foundation of their society in the 1980s. The drama of the oil shocks was slowly fading out of America's collective memory. Real oil prices were falling as new sources of supply, developed in the wake of the energy crisis, were flooding the market. The discourse of abundance was back. As David Nye observes, "By 1985 the belief in energy abundance through high technology was resurgent." He continues:

> Reagan's election was in part a public declaration of faith in the old order, in which high energy consumption went hand in hand with the dream of personal success, as measured in ever larger shopping malls, expansive houses full of appliances, large cars, and exurban living.[84]

By the end of the 1980s, the aggregate national efficiency gains achieved in the wake of the energy crisis had been offset by increased driving, rising car registration numbers and rising per capita energy consumption. A 1989 article in *Time* summed it up nicely:

> As if the 1970s were only a bad dream, consumers have been content to step on the gas. Sales of light trucks and four-wheel-drive vehicles, which generally guzzle more fuel than autos, have set US sales records for four of the past five years.

The article blamed low energy prices for these developments. The author added, with more than a hint of astonishment, that, when consumers chose new cars, "mileage per gallon ranks only fifth or sixth among their priorities, after color and floor plan".[85]

But in the shadow of this seeming return to the normal assumptions about abundance and consumption in the hydrocarbon society, something had changed for good. The effects of these gradual shifts became increasingly visible during the 1990s. Irrespective of the return of oil abundance, the emergence of the environment as an important object of politics could not just be undone. Established American assumptions about the benefits of oil consumption increasingly clashed with rival representations of energy emphasizing the fragility of the global ecosystem or the dangerous effects of burning fossil fuels on the earth's atmosphere. In 1989, the *Exxon Valdez*, an oil tanker, ran aground in the Prince William Sound. The resulting spill, the second largest in American history, contaminated 1,600 kilometers (about 990 miles) of Alaska coastline with crude oil and underlined forcefully that energy and environment could no longer be separated.

Growing concerns about climate change provided another challenge to the cultural foundations of the hydrocarbon society. Increasing scientific evidence and public awareness about the adverse environmental effects of burning fossil fuels did much to problematize oil consumption and to challenge the idea that consumption was simply and unproblematically linked to a better material future. Climate change became an issue in US politics early on. In 1988, a series of Senate hearings transformed the greenhouse effect from a science topic into a policy issue almost overnight. In the same year, presidential candidate George H. W. Bush declared that as president he intended to take firm action against global warming. Climate change had reached the White House. In 1990, the International Panel on Climate Change published its first assessment report on global warming. Two years later, President Bush snr travelled to Rio de Janeiro to sign the United Nations Framework Convention on Climate Change (UNFCCC). In 1997, Bush's successor, Bill Clinton, signed the Kyoto Protocol, thereby committing the United States to achieving tangible reductions in carbon dioxide (CO^2) emissions. The Senate refused to ratify the treaty and it never came into effect in the United States, but the whole affair nevertheless showed that climate change was now firmly a part of the public political discourse.[86]

Another challenge was presented by the debate about "peak oil", that is the prediction that the world will soon exhaust its finite oil reserves. This notion rests on the ideas of Marion King Hubbert, which he used in the 1950s to predict—fairly accurately, as it turned out—the peak of American oil production in the 1970s.[87] Later, oil experts like Colin Campbell and Jean Laherrère began using Hubbert's methods to forecast the point at which global oil production was likely to peak.[88] Although there has been much controversy about the accuracy and usefulness of such peak oil forecasts, the specter of a world without oil has been looming large in the public debate ever since.[89] As Imre Szeman observes, predictions of peak oil are almost always oriented toward the "anticipation of disaster" that is associated with the end of oil. Modern Western lifestyle is so fundamentally dependent on petroleum, and oil does so much to define the very nature of modernity, that running out of it appears as a fundamental threat to human civilization on a par with the threat of climate change.[90]

Both the climate change and peak oil debates implied scarcities, imbalances and limits that made it increasingly hard to talk about oil as

America's benign lifeblood. Instead, this metaphor's main rival, the representation of oil as an addiction, moved closer to the center of American oil culture in the 1990s. Environmentally conscious liberals like Al Gore used it frequently. Arguing that global warming was quickly becoming "one of the most serious problems humans face", Gore insisted that it was high time to "confront our addiction to fossil fuels, especially foreign oil, and measure the full costs of the energy choices we make".[91] Even ardent conservatives began to speak about dependency in these terms. Charles Krauthammer called the United States "a nation of oil addicts" in the *Washington Post* and suggested that the country needed to reduce its oil reliance.[92]

And then came George W. Bush's 2006 State of the Union address. Bush declared that the United States was addicted to oil. As the brief discussion at the outset of this chapter shows, the speech single-handedly transformed the notion of an oil addiction into a buzzword used to describe society's relationship to oil. Addiction became a major, and perhaps the primary, representation of dependency in the United States. The shift from lifeblood to addiction reflected a gradual process of redefining dependency as a political problem. Increasingly, the security threats traditionally thought to result from import reliance were now associated with all forms of American oil dependence. In May 2006, a few months after Bush's speech, the US Senate Committee on Foreign Relations organized a hearing under the headline "Energy security and oil dependence." In the opening statement, Chairman Richard Lugar said: "The committee meets today to consider strategies for reducing dependence on oil. This dependence brings intolerable costs to American national security and economic well-being."[93] Even more explicitly, a redefinition of general oil dependence as the key energy problem facing the nation was evident in one of the testimonies that followed. Jason Grumet, executive director of the National Commission on Energy Policy, argued that oil was a fungible commodity traded in a single global market and that, therefore, "our economic vulnerability to oil price shocks is entirely a function of how much oil we use—the continent from which the oil was extracted has no bearing, whatsoever, on this equation". He said further, "If we accept that the key measure of our energy security is not how much oil we import, but how much our economy depends on oil, we can begin to articulate more realistic goals and actually set about achieving them."[94]

Barack Obama took another forceful step toward redefining dependency. He too frequently used the addiction metaphor, but also evoked a new image to make sense of America's increasingly troubled relationship with oil. Obama vowed to end "the tyranny of oil in our time".[95] Nothing could be further removed from the glittering assumptions that had once characterized the hydrocarbon society: oil as black gold, as a juice from the fountain of youth; oil as progress, mobility and freedom. The American people, the President claimed, had been enslaved by fossil fuels. The only solution was to build "a new energy economy", one constructed on a foundation of efficiency, clean energy sources and reduced greenhouse gas emissions. In other words, the country needed a third energy transition.[96]

The necessity to transition away from oil was further highlighted in the context of the Gulf of Mexico oil spill in 2010. For three months, gigantic amounts of oil spilled uncontrollably from BP's deep-water Macondo Prospect into the Gulf of Mexico, exposing the nation to a daily spectacle of oil-covered birds, contaminated shorelines and destroyed livelihoods. It was the greatest environmental accident in US history. Although not comparable to the 1973 embargo, the 2010 oil spill delivered a discursive shock of sorts to the energy foundations of American society. The crisis rendered even more precarious the assumption that increased domestic energy production was desirable, safe and necessary for national security or energy independence. From Dick Cheney's National Energy Policy plan in 2001 to Sarah Palin's "drill, baby, drill!" campaign in 2008, pressure to open up ecologically difficult areas—mainly parts of the outer continental shelf and Alaska—for domestic production had increased steadily in an environment of high energy prices.[97] Just weeks before the explosion of the *Deepwater Horizon* oil rig, Obama had relaxed his stance against offshore drilling, saying he was willing to allow some exploration of the outer continental shelf as part of a broader strategy to change the nation's energy mix toward sustainable and clean energy. In the official narrative that the President provided to the nation in the midst of the crisis, Obama firmly emphasized this transition theme. In a televised address to the nation, the President described the disaster as a lesson about "our addiction to fossil fuels". Obama explained:

> For decades, we have known the days of cheap and easily accessible oil were numbered. For decades, we've talked and talked about the need to end America's century-long addiction to fossil fuels. And for decades, we have failed to act with the sense of urgency that this challenge requires [. . .] And today, as we look to the Gulf, we see an entire way of life being threatened by a menacing cloud of black crude. We cannot consign our children to this future.[98]

In the wake of the spill, the necessity of an energy transition was proclaimed in various places in the United States. A number of books appeared explaining the need, prospects and implications for such a transition.[99] *Science* editor Donald Kennedy argued that it was time to end the nation's addiction to fossil fuels, which he called "the real axis of evil in the United States today".[100] Thomas Friedman wrote in the *New York Times*, "The only meaningful response to this man-made disaster is a man-made energy bill that would finally put in place an American clean-energy infrastructure that would set our country on a real, long-term path to ending our addiction to oil."[101] All of this had been said before, during and after the energy crisis of the 1970s. This time, however, it seems that Americans have begun to reframe their relationship to oil in earnest. The implications of this shift for American society, for the material and cultural foundation of life in the United States, are enormous. After all, a third energy transition would not simply change the hydrocarbon society it. It would destroy it.

Foreign Oil Dependence

Foreign oil dependence was imagined as the nation's key energy problem in the wake of the 1973 energy crisis, and this notion persisted long after memories of the oil shocks began to fade. Foreign oil dependence had acquired the status of common-sense knowledge, a threat that no longer had to be explained in detail, something Americans no longer needed to be convinced about. Most importantly, dependency remained an important feature of general political discourse even after energy ceased to be, in Yergin's words, "the hottest cauldron in national politics".[102] Owing to more favorable price and supply developments in the energy markets, public interest in energy gradually declined in the 1980s and 1990s.[103] Despite this, US presidents continued to talk about the dangers of foreign oil. Ronald Reagan issued a warning about the United States' "continued vulnerability" to oil supply disruptions.[104] George Bush snr called dependence on oil imports "excessive".[105] Bill Clinton declared, "The Nation's growing reliance on imports of crude oil and refined petroleum products threaten the Nation's security."[106]

When energy re-emerged as an urgent political issue in the early 2000s in the context of rising energy prices and the terrorist attacks of 11 September 2001, this discourse of dependency had not lost its purchase. Poll after poll demonstrates that the vast majority of US citizens are still concerned about disruptions to foreign oil supply in the present. In 2007, to take just one example, 93 percent of respondents said that US foreign oil dependence poses a "serious problem" to the country, 70 percent even called it "very serious".[107] In 2013, another poll yielded an even more striking result. According to this survey, 90 percent of Americans deemed it important to try to reduce the country's reliance on oil from outside North America. And 63 percent of those polled considered this goal more important than reducing greenhouse gas emissions.[108] To these 63 percent, in other words, the dangers of foreign oil were greater, more urgent or more real than the dangers inherent in climate change. Despite the increasing emphasis on links between oil and environmental threats, this seems to suggest, the problem of dependency can still be imagined primarily as an issue of US foreign policy.

The power of this particular definition of dependency is further underlined by the continued popularity of energy independence as a goal of US energy policy. Since the 1970s, almost any presidential campaign and a majority of State of the Union addresses have featured promises of energy independence.[109] Such proposals often ignored what many energy experts have been saying: that true independence is impossible even if imports are reduced to zero. Oil, this argument goes, is traded on a global market today, a market in which price developments anywhere ultimately filter through to all market participants. "There is only one oil market", argues Daniel Yergin. "Secession is not an option."[110] Up until a few years ago, promises of energy independence also contrasted sharply with a seemingly unstoppable rise in US oil imports—from about 35 percent of domestic oil consumption on the eve of the 1973 embargo to more than 60 percent in 2005. Many energy professionals therefore tended to think of energy independence proposals as mere political rhetoric incompatible with the material realities of the oil market. To them, the longevity of energy independence was both a puzzle and a source of frustration. "Is there anything more depressing than yet another promise of energy independence in yet another State of the Union address?", sighed Charles Krauthammer in 2007.[111] And John Hofmeister, former CEO of Shell, asked, "What is it about 'energy independence' that keeps this mantra in the political rhetoric of political campaign after campaign?"

From the perspective on energy adopted in this book, the answer to Hofmeister's question is fairly clear. What has kept energy independence in the debate is the continued dominance of a definition of dependency that equates foreign oil with a number of severe threats to the nation. The promise of securing the United States against these perils is the flip side of the same coin—the culturally constructed solution to a culturally constructed problem. In this view, a policy of energy independence does not need to conform to the realities of the global oil market in order to be effective. The notion functions as reasonable, logical, legitimate and desirable within the logic of dependency that Americans have adopted after the 1970s. To call for energy independence is to affirm that the United States will remain independent, remain exceptional and remain free even in the face of foreign threats. On an emotional level, Hofmeister is well aware of this. As he writes:

> Well, even I have to admit: it sounds good! What message carries more bravado and more assertive masculinity than "energy independence"? It says, "We can tell the rest of the world to go to hell." There's something earthy, powerful, atavistic, and pugilistic, even legitimately xenophobic, about saying it. It speaks for all Americans regardless of gender, ethnicity, or age. It creates an image of national solidarity and the idea of a final one-upsmanship over the Organization of Petroleum Exporting Countries."[112]

Yet another sign that the discourse of dependency is still powerful today is that the current American energy boom has frequently been imagined in the terms provided by the twin notions of foreign oil dependence and energy independence. The revolution in shale gas and oil has been like "manna from heaven" for those who want to see the United States self-sufficient in energy.[113] Because of increased production and decreasing oil consumption, North America as a whole may be energy independent by the end of this decade. Even the

United States alone may come close to, or even achieve, self-sufficiency. The EIA predicts that by 2040 US dependence on foreign oil may still be about 32 percent of domestic demand. Under favorable circumstances, however, the country may be able to reduce its oil imports much more or even become a net exporter of oil.[114] In the area of refined petroleum products, the United States has already become a net exporter—a status the country has not been able to claim since the days of President Truman.

In the context of these developments, the possibility of energy independence has been celebrated in the United States as a major achievement of the country in the twenty-first century and a comprehensive solution to the problem of foreign oil dependence. "To not be concerned with where our oil is going to come from is probably the biggest home run for the country in a hundred years", as one oil industry executive puts it.[115] "Energy independence isn't just a pipe dream", wrote *USA Today*.[116] Energy economists continue to point out that the fungibility of oil makes true independence impossible.[117] But a significant amount of attention is now also directed towards the various benefits that energy self-sufficiency might hold for the United States. In their report *Energy 2020: Independence Day*, a team of analysts around Edward Morse finds that current developments will yield tangible national security gains for the United States in addition to a better current account balance, a stronger dollar and partial insulation from international price spikes. The United States, they argue, would no longer have to sacrifice a values-driven foreign policy promoting human rights and democracy in order to ensure good relations with resource-rich despotic regimes.[118] These arguments show that the perspective adopted to assess the implications of the American energy revolution is one still heavily influenced by the two Big Ideas of foreign oil dependence and energy independence. In such assessments, there is no trace of an alternative view depicting dependency as an addiction to which the country needs to respond by shifting away from fossil fuels. The way in which the United States has given meaning to its current energy boom demonstrates the continued power of a definition of dependency that centers on the dangers of foreign oil.

This leads to a final point: current American debates about dependency are heavy with ambiguity. On the one hand, the practice of consuming oil is increasingly being problematized in the United States today. On the other hand, there is a still powerful reflex to think about oil in the more narrow terms of foreign oil dependence and energy independence. In effect, a fundamental ambiguity has crept into political debates about energy issues, such that it is not always easy to understand what statements about dependency imply exactly. Reconsider, for example, Bush's declaration of dependence in his 2006 State of the Union address with which this chapter opened: "America is addicted to oil, which is often imported from unstable parts of the world." What is the key problem here, according to Bush? America's excessive reliance on oil as its key energy source? Or the fact that this oil has to be imported from elsewhere? Both readings are possible, and both appeared frequently in Bush's public statements about energy. For example, while Bush's addiction claim has often been taken to be a statement about foreign oil dependence, Bush later said it was not. As he clarified:

> I kind of startled my country when, in my State of the Union, I said, "We're hooked on oil, and we need to get off oil." That seemed counterintuitive for some people to hear a Texan say. But the truth of the matter is, we got to diversify away from oil.[119]

Hence, it had been a statement about excessive American oil consumption. On another occasion, however, Bush claimed it had been about foreign oil supply:

> I'm going to work with Congress to reduce our dependence on foreign oil. I know it came as a surprise to some of you that I would stand up in front of the Congress and say, "We got a problem; we're addicted to oil." But it is a problem. And I look forward to working with both Republicans and Democrats to advance an agenda that will make us less dependent on foreign oil.[120]

In much the same way, Barack Obama has mixed concern over the problem of foreign oil dependence with claims that the key US energy issue was its insatiable appetite for fossil fuels in general. This indeterminacy is evident, for example, in Obama's repeated calls to end America's "addiction to foreign oil", a choice of words that blends a body-politic metaphor representing dependency as a problem of domestic demand with the more traditional focus on foreign supply.[121] Sometimes, foreign oil dependence is depicted as a consequence of American oil consumption. Sometimes, a transition away from oil appears as necessary in order to escape the dangers of foreign oil. In both cases, however, foreign oil is associated with danger. At the present juncture, American debates about dependency contain two definitions of oil dependence as political problems and a meaningful distinction between these modes of representation has begun to break down.

Conclusion

Dependency has been the principal feature of the hydrocarbon society since at least the 1920s. The genealogy of dependency presented in this chapter demonstrates an important point that is not usually made explicitly in accounts of US energy policy. There has been no necessary, simple or stable correspondence between dependency as a thin material fact and definitions of dependency as a political problem. As a discursive construction, dependency has been fluid, contingent, incoherent and contradictory. It has been created by policy makers, energy and foreign policy experts, environmentalists, activists, thinkers and the news media out of the extant cultural raw materials contained in oil's discursive economy in the United States—including deeply held assumptions about abundance, consumption, progress, exceptionalism and freedom. Nowhere did the cultural and contingent nature of these definitions of dependency become more evident than in the wake of the discursive shock of the first American energy crisis. The crisis opened up space for entirely new ways of imagining oil in a society rocked by anger, confusion and uncertainty. In the context of the oil embargo against the United States, US officials created a narrative according to which the energy crisis was a problem of ensuring foreign supply and not one of excessive domestic demand. The oil shock, in this view, was the consequence of the country's dependence on foreign oil and the reckless acts of a few Arab oil-producing countries. The association between foreign oil and existential threats at the heart of this interpretation was then carried forward by every administration following Nixon's. It permeated American political debates far beyond the Washington policy environment, crowded out alternative representations and almost acquired the status of self-evident truth.

Moreover, this reading of the events also set the direction for US energy policy. Most importantly, the principal solution suggested by the dominant definition of dependency was to ensure supply—by means of energy independence, or some version of it; diversification of foreign supply away from OPEC; promoting domestic and non-OPEC exploration; policing the Middle East; and, if there was no other way, deploying the US military to keep the oil of the Persian Gulf flowing. The alternative of reducing domestic demand was often debated but not implemented with real consistency.[122] As Richard Vietor argues, "Demand-reduction ran counter to the values of a people accustomed to abundance, and served few of the organized constituencies that traditionally influenced public policy."[123] This further underlines how energy culture in the hydrocarbon society has enabled some and foreclosed other courses of action in US energy policy. In a 1975 congressional hearing, Treasury Secretary Simon made the telling remark that rationing and other policy interventions in the basic freedoms to consume would be unacceptable to the American public: "We could perhaps live with rationing in a period of temporary emergency. But as a way of life, I suggest it is fundamentally inconsistent with our system and with the spirit of the American public."[124]

CHAPTER 3

THE DANGERS OF FOREIGN OIL

In August 1975, the Congressional Research Service transmitted a report to the Committee on International Relations of the House of Representatives. The subject of the report was potentially explosive: would the US military be able to seize and control the oilfields of the Middle East to ensure adequate oil supply in the case of a second oil embargo? Democratic Congressman Lee Hamilton had commissioned the study in response to public claims by top government officials, including President Ford, that US foreign oil dependence was so dangerous that another embargo might push the country to the precipice of collapse.[1]

The analysts at the Congressional Research Service began their assessment by outlining the exact nature and scope of the threat posed by US reliance on oil imports. They noted that oil provided almost half of the energy consumed in the country, and that 27 percent of this oil currently had to be imported. The nation, in effect, was severely susceptible to oil embargoes and other deliberate or accidental cut-offs of petroleum supply. They also stressed, however, that "not even a full-scale OPEC oil embargo would threaten US survival, our only vital interest". Unlike Europe or Japan, the United States would be able to adjust permanently to global shortages of oil. While possible, however, such an adjustment would be neither easy nor painless. "In the process", the authors predicted, "severe economic and social problems probably would alter America's fundamental lifestyle."[2]

This report is a valuable source for the purpose of this chapter because it allows us to grasp what is at stake when it comes to dependency and, just as importantly, what is not at stake. The notion of threat is of central importance in this regard. The previous chapter, by tracing how Americans have historically defined dependency, described how foreign oil dependence first emerged as a serious problem in the course of the energy crisis of the 1970s. The present chapter turns to examining the central claim underpinning this definition of dependency: that US reliance on large quantities of imported petroleum threatens the national security of the United States. The articulation of causal links between dependency and threat was part and parcel of narrating the energy crisis in the 1970s and subsequently came to be accepted almost as a self-evident, taken-for-granted description of the world. US presidents and other state officials after the 1970s could simply refer to the need to reduce US oil imports or to achieve energy independence, without much further explanation. Today, Barack Obama can simply state, without providing any details or context for this claim, that "our economy, and our Nation itself" are endangered by foreign oil dependence, and this will be recognized by an American audience as a meaningful and reasonable statement.[3] To say "foreign oil" is enough to invoke an array of vague but seemingly severe threats to national security, firmly associated with dependency.

This chapter identifies and structures the assorted dangers of foreign oil as they have appeared primarily in the statements of policy makers and professional observers of US foreign policy since the 1970s. The goal here, however, is not to assess the empirical accuracy of such claims or to discern how serious oil-related threats to national security really are.[4] Instead, this chapter asks a different set of questions: What has it meant to say that foreign oil dependence threatens the national security of the United States? What are the dangers posed by foreign oil? What, precisely, is the object of these threats? How have they come into existence?

The conceptual foundation for these questions is provided by the insight that threats are not self-evident facts but cultural constructions. This is not to say, of course, that there are no dangers in the world, but rather that there is no necessary correspondence between the risk inherent in some event and its representation as dangerous in a given society.[5] Jutta Weldes argues that US foreign policy threats emerge as part of a process by which representatives of the State and other important actors assign meaning to the world. Before national interests can be defined and concrete policies formulated, these speakers populate the international system with actors and objects as well as relationships between them. This process involves assigning identities to the principal acting subjects, including the United States, which are described in relation to the threats and national interests at hand. In effect, "identity and insecurity are produced in a mutually constitutive process".[6]

The relationships between the objects and subjects constructed in this manner, Weldes says, "often appear in the form of quasi-causal arguments". They are "quasi-causal" rather than causal links because "the relations and causal chains they posit may or may not be empirically valid on their own terms". Irrespective of their factual accuracy, the meanings created in this process matter because they render certain beliefs, political positions or courses of actions reasonable, legitimate and justified. As Weldes puts it, "They help to specify among other things which objects are to be protected and which constitute threats."[7]

These causal links between dependency and threats to US national security were forged in the political debate about the energy crisis in the 1970s. US officials, the news media and policy experts explained that dependency caused economic imbalances and, in case of an interruption of supply, economic recession. Dependency caused the deterioration of the high living standards enjoyed by most Americans. Dependency caused potentially devastating foreign policy dilemmas. It caused constraints on the sovereignty and independence of the United States and the decline of American power. Dependency caused vulnerabilities that the enemies of the United States could use to hurt the nation. These views shape American debates about energy to the present day.

Examining such claims about the dangers of foreign oil amounts to a search for the object of energy security. What is it that is endangered by dependency? What is it that needs to be protected against the threats arising from import reliance? What, to borrow a term from theories of securitization, is the "referent object" of energy security?[8] This chapter argues that the principal object of energy security as constructed in the discourse of dependency is national identity. That is to say, US foreign oil dependence threatens the viability of certain material and representational practices that are central to American images of the essential nature of the nation. Ultimately, the central anxiety that runs through the American discourse of dependency is that "we" might no longer be able to be "us" because of dependence on foreign oil.

The Congressional Research Service report quoted above offers a fitting entry point into this body of representations. Its authors clearly state that the physical survival of the United States is not at stake. Even the most comprehensive oil embargo would not threaten the military security or the territorial integrity of the United States. What is more, despite frequent observations in the academic literature about the central importance of oil to the modern military machine, US dependence on foreign oil has not usually been described by policy makers or security professionals as a problem of ensuring the capabilities of the US military. In 1970, for instance, the government task force on oil imports and national security concluded that petroleum imports "do not, in the main, concern any danger to the functioning of the nation's armed forces".[9] Today, US foreign oil dependence tends to be portrayed as a significant burden on the military budget or as a logistical issue but not necessarily as a fundamental problem threatening the US military's ability to defend the country.[10] The dangers of foreign oil are thus not primarily located within a classical realist conceptualization of national security as securing the continued existence of the state against existential foreign threats.

The real threat posed by a foreign interruption of the US oil supply, according to the experts at the Congressional Research Service, was "economic" and "social" and concerned the risk that a long-term interruption would change "America's fundamental lifestyle". Nixon's task force arrived at a very similar conclusion. The major risk associated with foreign oil dependence was an interruption of "essential civilian demand" for things like motor fuels, home heating and industrial lubrication, which would have "serious effects on the nation in general".[11] It was precisely this effect on "the nation in general" which turned oil imports into a major threat in the 1970s. What was at stake here was not the survival of the United States but the survival of the country *as a hydrocarbon society*. That is to say, dependency threatened the viability of particular modes of organizing everyday life in the United States, the ensemble of practices that constitute the American way of life. Echoes of these fears resonate within US energy discourse until the present day.

The argument that energy security is about securing the American way of life is sometimes put forward in books about oil, but usually in an unnecessarily narrow sense. Michael Klare argues that oil is so central to the American way of life that this lifestyle could not survive a large-scale, continued cut-off of oil supply.[12] Ian Rutledge notes that the concept of energy security in the United States refers primarily to the preservation of an American way of life based on oil-fuelled automobility.[13] Bruce Beaubouef observes that American policy makers over time determined that the protection of established patterns of energy consumption, and thus the protection of American energy consumers, was an integral part of national security.[14] These statements already contain within them the notion that the ultimate object of energy security is American identity as encapsulated in a specifically American way of life. They tend to be confined, however, to material aspects of energy use that characterize American society. This is an unnecessary limitation.

This chapter also argues that there is a link between oil-related threats, national security and American identity, but it expands and deepens this argument in three ways. First of all, the threats formulated in the discourse of dependency concern not only the sustainability of material patterns of energy consumption in the United States but also the viability of a number of cultural discourses that have complemented and supported these energy practices in the hydrocarbon society. The previous chapter already showed that energy crisis was widely described as a threat to key building blocks of energy discourse in the United States, including discourses about abundance and progress. In addition to these challenges, foreign oil dependence was also portrayed as a threat to the American dream of individual opportunity, social mobility and the promise of a better future.

Secondly, US foreign oil dependence also became a threat to the security of established descriptions of the nation's founding values and the principles on which US foreign policy was based. Some of these themes are not inherently related to petroleum, and go back beyond the beginnings of the hydrocarbon society. In particular, dependency contained an implicit but severe threat to the time-honored notion of American exceptionalism. The discourse of exceptionalism revolves around the notion that the United States is somehow exempt from "the universal tendencies of history, the 'normal' fate of nations, the laws of historical mechanics itself".[15] Under the big umbrella of exceptionalism, US policy makers have traditionally formulated various goals, ambitions, principles and values that defined the place and the purpose of the United States in the wider world.[16] Within the discourse of dependency, the necessity to import oil to sustain the level of US energy consumption was represented as a severe blow to some aspects of exceptionalism: independence as a founding principle of the nation; the role of the country as a great power, and the leader of the Western world; and the notion that the national interests on which US foreign policy were based conformed to high moral standards, not just material necessity.

Thirdly, dependency should not simply be seen as a threat to descriptions of American identity in the way described above. The issue is more complex. David Campbell has argued that foreign policy is a meaning-making activity deeply implicated in the affirmation and maintenance of the identity of the State. The articulation of threats to national security in foreign policy discourse plays an important role in this regard.[17] Foreign policy, in Campbell's view, is a "discourse of danger" that serves to maintain and reinforce that which it describes as being threatened. "The constant articulation of danger through foreign policy is thus not a threat to a state's identity or existence; it is its condition of possibility."[18] In this view, speaking about the dangers of foreign oil has afforded US policy makers and others involved in the discourse of dependency an opportunity to affirm and reinforce American descriptions of Self.

The first part of this chapter traces the original formulation of threats related to foreign oil as they appeared in the 1970s. It focuses on three key threats: dependency as a threat to the American way of life, dependency as a threat to American independence and dependency as a threat to American international leadership. The second part of the chapter examines how the content and logic of this discourse evolved after the 1970s, focusing on US energy debates during the presidency of George W. Bush. In particular, it presents how these threats have been re-imagined in the context of concerns over international terrorism after 11 September 2001.

Foreign Oil and Threat in the 1970s

The link between oil imports and threats to national security was forged in the context of the 1973 Arab oil embargo against the United States and the energy crises of the 1970s more generally. This does not mean, it bears repeating, that Washington policy makers had no concerns over foreign oil prior to that time. What it does mean is that in the 1970s the dangers of foreign oil acquired the character of a self-evident, taken-for-granted description of the world in political debates well beyond narrow government circles.

Before the embargo, many observers in the United States did not take the claim that dependency exposed the nation to significant risks very seriously. In 1970, President Nixon's cabinet-level task force on oil import control concluded that the risk of potentially dangerous embargoes was so low that oil import controls could be partially relaxed.[19] Even when Arab states explicitly warned the US Government that oil could be used as a weapon under certain circumstances, few people in Washington paid attention. In the run-up to the Yom Kippur War in 1973, Saudi Arabia had issued several clear warnings that US oil supply would be affected in the case of a military conflict; to no avail. After all, the Arab producers had tried it twice before: during the Suez Crisis in 1956 and the Arab—Israeli War of 1967. In neither case had the embargoed countries experienced any significant supply shortages.[20] On 19 October 1973, just one day before the embargo took effect, the editors of the *Wall Street Journal* claimed with astonishing confidence that the prospect of an Arab embargo against the United States "isn't exactly the most devastating threat in the history of economic warfare".[21] Only days earlier, the same journalists had already called what they saw as an "oil bluff". In their view, it was "ludicrous to believe that the United States [...] should in the long run have to worry about a secure energy supply".[22]

Such doubts about the dangerousness of the oil weapon were quickly wiped out, however, when the combined effects of the embargo and heavy-handed government policies to deal with the situation began to produce visible effects in the US economy. Suddenly, fears about the dangers of foreign oil mushroomed in policy circles and all over the national news. In the midst of skyrocketing energy prices, long gas lines, nationwide rationing of fuel, inflation and rising unemployment, the national debate about energy took on a "paranoid style".[23] The suddenness with which this apocalyptic tone had appeared was widely noted. Robert Tucker, a Washington-based professor of international relations, noted that even those who seemed to be prophets of doom before the crisis "in retrospect appear as unguarded optimists".[24] *Time* observed that "cataclysmic predictions" circulated in Washington, ideas that would have been dismissed as ridiculous just one year earlier.[25] "Otherwise sober men spoke of extreme solutions: repudiation of international debts, massive currency devaluations, the suspension of parliamentary government, even military intervention in the producing countries."[26] According to the US ambassador in Saudi Arabia, James Akins, even Washington's instructions to Riyadh during the crisis "bordered on hysteria".[27] A sense of impending doom also gripped the American public. A reprint of John Walvoord's book *Armageddon, Oil, and the Middle East*, a Bible exegesis that predicted the end of the world over a Middle Eastern oil crisis, sold more than 600,000 copies.[28]

The second oil shock at the end of the 1970s only served to reinforce this sense that foreign oil dependence threatened the nation's future. The oil-supply disruption resulting from the Iranian Revolution was, as the *Los Angeles Times* put it, a "disquieting exercise in deja vu" which "raised once again a nagging set of questions about the security of the nation's access to that most critical of all commodities: energy".[29]

These questions were part and parcel of a much broader debate about the end of the country's international supremacy and the decline of American power. Yet, the energy crisis was not the only event that contributed to this debate. The Watergate Scandal and the catastrophic military failure in Vietnam also fuelled a widespread sense that American power was waning.[30] "In America, there is the now fashionable theme", wrote Robert Tucker, "that we have reached the outer edges of our power and that we must learn to accommodate ourselves to events and trends we may comprehend but cannot alter."[31] The drama of the energy crisis, the loss of American self-sufficiency in oil and the emergence of the discourse of dependency helped shape this notion of decline. Looking back at the events in 1985, Stephen Krasner argued, "The decline of American power and changes in a postwar economic regime are nowhere more dramatically evident than in the area of crude petroleum."[32]

Foreign Oil and the American Way of Life

One central link between US foreign oil dependence and national security concerned the domestic economy of the United States. It became routine for policy makers, economists and journalists to point out that dependency threatened the high living standards enjoyed by most Americans. This threat was said to result from the central role of energy supply for the continued growth of the economy. As Gerald Ford put it in May 1975, "Our American economy runs on energy—no energy, no jobs. In the long run, it is just that simple."[33] The dominance of petroleum in the US energy system meant that oil was not just a simple commodity, it was *the* central commodity for the United States and an essential precondition for growth and material progress in the country.[34]

In January 1977, President Ford sent a special message on energy to Congress in which he explained at length that the dangers of foreign oil were located primarily in the field of the domestic economy. The text contained a comprehensive summary of the country's energy situation as the outgoing Ford Administration saw it. The President began by declaring that the central energy problem facing the nation was dependence on foreign oil. Then, under the headline "The costs of dependence", Ford explained in great detail why this was a problem. Since it is typical of countless similar descriptions of the economic dimension of dependency, the President's statement is worth quoting at length:

> The real price paid for our growing dependence on imported oil is our vulnerability to significant economic and social disruption from the interruption of oil imports. Apart from the inconvenience experienced by millions of people, the 1973–74 embargo and the resulting higher prices caused a loss of about 500,000 jobs and approximately $20 billion in our Gross National Product. The sudden four-fold increase in OPEC oil prices contributed significantly to inflation. Since 1974 our dependence on imports, particularly from Arab nations, has grown by a million barrels per day, so that an interruption of supply today would be even more disruptive of our economy than the 1973–74 embargo. Another cost of energy dependence is the outflow of US dollars to pay for imported oil, totalling about $34 billion in 1976 or $160 for each American, eleven times that in 1972.[35]

The economic dimension of the threat posed by oil-import reliance is so important because, within the discourse of dependency, it was translated into an existential threat to the foundations of the hydrocarbon society. It has been argued that the economic facet of energy security is fundamentally different from more traditional concepts of national security because what is at stake here is not the survival of the state but merely the smooth operation of the economy. As one scholar puts it, "The need for energy is not driven by the imperative to survive, but by the functional demands of various sectors of activity, which means that its absence does not lead to extinction, but to dysfunction."[36]

It is certainly correct that threats to the operation of the economy do not fall within the scope of narrow definitions of national security, understood as military security and the absence of existential threats to the territorial integrity of the state. However, this does not mean that economic threats associated with energy cannot be existential. In the discourse of dependency, US policy makers and other observers routinely claimed that foreign oil dependence undermined the viability of the American way of life. What was said to be at stake, in other words, was not the survival of the United States as a sovereign state but the survival of American society in its current form. The hydrocarbon society had been structured around abundant and cheap supply of petroleum to such an extent that disruptions of the inflow of foreign oil amounted to a full-frontal assault on its material and cultural foundations.

A close reading of statements produced within the policy community and the national news media in the wake of the embargo clearly shows that the dangers of dependency were seen to far exceed the mere threat of economic dysfunction. The *Los Angeles Times* warned that the national economy was at risk of becoming dependent on decisions made in the Middle East. If American dependency increased as projected, the newspaper feared, the Arab producers would "have the power to paralyze the US economy, creating mass unemployment, anytime they decided for whatever reason to cut off oil supplies".[37] Two further examples underline the severity of this threat. In 1976, Ford's vice-president, Nelson Rockefeller, appeared before a Senate committee to explain the administration's energy policy. Noting the extent to which the country now depended on oil imports, the Vice-president asserted that the stakes were extremely high: "If this supply were suddenly cut off, there would be social and economic chaos. Should we have another embargo, the economy of this country would be shattered." A future boycott, Rockefeller continued, could mean nothing less than "total breakdown". Inactivity in the face of rising oil imports meant risking "the future of our survival as a society".[38]

The second example is a *Washington Post* editorial from February 1974, which described how the effects of the oil embargo undermined the daily routines of life in the United States:

> The lengthening lines at the filling station are already far more than a mere passing nuisance. They are rapidly becoming a real affliction and, for many people, a threat to their livelihoods. It is all very well for government officials to repeat the sociological cliches about our wasteful styles of life. But that lofty perspective offers cold comfort to the family living out in the suburban subdivisions that federal housing policy has assiduously encouraged for a quarter of a century. There the immediate necessity is getting to work, getting to the store, and getting the children to their dentist's appointments. For a great many jobholders [...] there is no possibility of public transportation to work. For many businesses too small to buy fuel in bulk, it is a matter of lining up for gas in order to stay in operation.[39]

The editorial made it abundantly clear what dependency potentially meant. The hydrocarbon society had been constructed as a system that forced Americans to consume oil for the most basic everyday activities. If supply of this oil was withheld for long enough, even affluent Americans would see their lifestyles changed beyond recognition. The principal victims of the crisis were thus the people most immersed in the US energy system: the American middle-class family.

Another way of demonstrating the existential nature of this threat is by studying the body-politic metaphors that many speakers used to describe the dangers of dependency. The problem, as one observer described it, was not that the American economy was going to suffer from a "heart attack", that is sudden death through embargo. The real problem was the "creeping emphysema" of the body politic due to its excessive dependence on foreign oil.[40] Another writer feared that dependency might "suffocate the US economy", and suggested a policy of energy independence as "the only route to survival".[41] Yet another described the United States as a "generally sick man", "haemorrhaging" and "bleeding to death". There was a "distinct possibility of an economic and political disaster bearing more than a superficial resemblance to the disaster of the 1930's".[42] What all these examples show quite clearly is that the dangers of foreign oil dependence were seen as exceeding mere economic dysfunction. Encapsulated in socio-medical terms, the risk of the death of the nation was invoked routinely and explicitly.

Ultimately, what was imperiled by foreign oil dependence was not simply the domestic economy but the viability of the material and representational practices that constituted the United States as a hydrocarbon society. There were several ways in which economic dysfunction translated into this much more existential threat. Most importantly, disruptions of economic activity threatened two central notions around which American identity has been imagined: equal opportunity and social mobility.[43] As Chapter 1 described, oil has always been closely linked with social mobility and the promise of a better future in the hydrocarbon society. As "black gold", it functioned as a symbol of progress and wealth in both political and literary energy narratives. In the context of foreign oil dependence, however, oil no longer enjoyed the status of an unambiguously positive symbol of individual opportunity and limitless material progress. Petroleum was still widely described as a catalyst for modernization and wealth, but this time it was Arab progress and Arab wealth. Even worse, oil symbolized Arab progress at the expense of Americans living standards. Consider, for example, the following account of the consequences of the energy crisis on the United States:

> These days, Americans are cramping themselves into compact cars to save on gasoline as it approaches the dollar-a-gallon level. In Saudi Arabia, meanwhile, the sons of camel drivers ride the streets of Riyadh in Cadillacs [. . .] Our cities are decaying and struggling to stave off bankruptcy. In 1960, Abu Dhabi was a dry fortress and fishing village. Today, because of the oil billions pouring into the United Arab Emirates, it is a gleaming metropolis of luxury hotels, boulevards, marble mansions and shopping malls.[44]

There were many similar stories in national newspapers during the 1970s, emphasizing the tremendous and sudden shift of material wealth away from the United States and into countries that had so far been considered economically marginal, backward and insignificant. Oil, in effect, turned into a more complex and ambiguous symbol of (foreign) wealth and (American) decay. Politicians and economists argued that high oil prices set by OPEC amounted to a tax on American consumers, which allowed the Arab producers to modernize but simultaneously threatened the health of the US economy. Economists like Walt Rostow argued that postwar economic growth, which he described as "the most remarkable phase of economic expansion the world has experienced in the last 200 years", had rested on two pillars. Both of them were connected to oil: the diffusion of durable consumer goods—above all, the automobile—and the emergence of a suburban mode of life. The stagflation haunting the country now, Rostow explained, had "essentially" been caused by high energy prices. He feared that they were "corroding the social and political life" in the United States.[45]

There was a widespread sentiment in the policy community that, beyond the temporary crisis triggered by the oil shock, something had been lost forever. The problem of dependency, in effect, signaled the end of an era. As Albert Ullman, chairman of the House Committee on Ways and Means, declared at the outset of a series of congressional hearings about the energy crisis, "There is no question that this Nation must make some hard decisions in the field of energy that are going to greatly change our traditional mode of living."[46] Such fears frequently found expression in expectations that the United States would have to adapt to a new reality, and that this adjustment would involve "sacrifices" and "painful change". Consider this *Los Angeles Times* editorial:

> Adapting to the new energy era is requiring some painful changes in the daily lives of Americans, changes whose relation to the nation's energy balance is rarely noticed by the average citizen. The changes include increased inflation and slower rates of economic growth, which in turn translate into reduced purchasing power and poorer job prospects for many households. Moreover, many analysts believe the pain is likely to worsen in the future. "Let's face it," an oil company executive says in a moment of candor, "Not all of our children will get as much of the good things of life as we feel we're entitled to".[47]

In this reading of dependency, the American middle-class family appeared as the principal social site where economic dysfunction translated into a challenge for the material practices and social relations that constituted the American way of life. As Natasha Zaretsky argues in her study on the American family during the 1970s, a peculiar effect of the oil embargo seemed to be its direct encroachment on the private sphere of ordinary Americans: "In the weeks and months that followed [the embargo], there emerged a frightening if embellished picture of imperiled middle class domesticity that suggested that the material comforts of American life were under direct attack." Rising unemployment and record peacetime inflation, in particular, seemed to chip away not only at established standards of living but also at traditional gender images and middle-class identities. The embargo shed a light on the inability of unemployed men to fulfill their function as breadwinners for the family, while the rise of the two-earner family further challenged the traditional social roles of women. The tumultuous reversals of authority that seemed to have occurred between the petroleum-consuming and-producing nations, Zaretsky argues, found an echo within the average American middle-class home.[48]

In sum, the economic dimension of the threat posed by the nation's dependence on imported oil was represented in the political debate as far more than a temporary problem of economic management. It became instead an existential threat to the future of the country. In particular, dependency seemed to undermine the living standards of the American middle class and the promise of a better material future through economic growth. Making explicit what often remained unstated, Gerald Ford claimed that "Americans are no longer in full control of their own national destiny when that destiny depends on uncertain foreign fuel at high prices fixed by others."[49] In much the same way, Ronald Reagan later observed that the energy crisis had turned "the greatest country on Earth into a land of broken dreams".[50]

Foreign Oil and American Independence

The second major articulation of threat to emerge from the energy crisis was the notion that dependency undermined US political independence and sovereignty—understood here as the ability of the government to pursue a policy of its own choosing in its international affairs. The embargo seemed to prove that any action counteracting the interests of the oil-producing states could be punished immediately and directly by the producers. As one journalist argued, "The real American stake in the energy crisis [...] is nothing more nor less than the political and strategic independence of the United States."[51] Since the stated goal of the OAPEC embargo was to change US foreign policy in the Middle East, US policy makers routinely used terms such as "coercion" or "blackmail" to describe the events. The US Secretary of State, Henry

Kissinger, argued that the embargo amounted to the "manipulation of raw material supplies in order to prescribe the foreign policies of importing countries".[52] Oil had become an instrument in the hands of the producers to direct the course of US foreign policy.

Central to this claim was the metaphorical representation of oil as a weapon wielded by the producers. In Arab political discourse, oil had been discussed in those terms since the 1950s in relation to the ongoing conflict with Israel.[53] However, after two largely unsuccessful Arab attempts to use oil as a political lever, in 1956 and 1967, it was only really in 1973 that the notion of oil as a weapon also entered the consciousness of the wider American public. Oil became, as one US senator put it, "an economic dagger at the throat of this country".[54]

The use of metaphors describing oil as a weapon was one way in which US speakers constructed the Arab oil producers as powerful actors. At the same time, the oil-importing countries were often imagined as utterly helpless before the supreme might of the producers. It was precisely this gap in power that made the oil weapon so dangerous and, ultimately, constrained the United States to conform to demands imposed on the country from outside. This claim of American powerlessness was a widespread, if often unspoken, element in US policy debates about the energy crisis. It derived from a rather pessimistic interpretation of the relationship of energy interdependence, or mutual dependence, between the exporters and importers of energy. As the concept of mutual dependence implies, US dependence on foreign supplies of oil was mirrored by the dependence of the producers on Western oil demand. In many cases the Persian Gulf states—most importantly, Iran and Saudi Arabia—also had close political, economic and military ties to the United States. The producers could use their natural resources as a weapon only at the risk of severing these ties.[55]

Prior to the embargo, American policy debates frequently contained claims that the country's foreign oil dependence formed part of a more or less balanced symmetry of dependency between the United States and the Middle Eastern producers. The United States needed foreign supply to keep the wheels of the economy turning, but the producers also needed foreign demand to receive revenue from the development of their oil resources. The Arabs cannot drink their oil, it was often said, and so they have to sell it.[56] Even if they were inclined to use oil as a weapon for political reasons, an embargo would pose, as the *Wall Street Journal* put it, the "danger of forever cutting off Arab access to the lush American energy market".[57] A few weeks before the Yom Kippur War, President Nixon had explained the situation in these terms at a White House press conference: "Oil without a market [. . .] does not do a country much good. We and Europe are the market." If Arab leaders behaved irresponsibly, Nixon continued, "the inevitable result is that they will lose their markets, and other sources will be developed".[58]

In the turmoil that followed the onset of the oil embargo in October 1973, these views on the nature of US—Middle Eastern energy interdependence shifted dramatically. While some observers still described interdependence as a roughly symmetrical relationship, these arguments were now up against an increasingly dominant interpretation that emphasized American powerlessness in the face of dependency. To say that the United States was living "at the mercy of the Arabs" became commonplace. Even Richard Nixon and Gerald Ford used the expression.[59] Journalists, academics, policy makers and ordinary Americans found a multitude of ways in which to express this pervasive feeling of helplessness, loss of independence and vulnerability. This sentiment also mixed with a sense of incredulity that a military superpower like the United States could find itself in such a position of weakness. Jack Anderson, a well-known journalist, lamented "how the United States has been reduced to petroleum dependency on a group of Arab rulers only a generation removed from tribal feudalism". Through a tremendous shift of power toward the oil producers, the nation had become an "oil puppet", mechanically following the orders of the Middle Eastern "oil puppeteers".[60] Others sometimes invoked the image of Dr Pavlov and his dogs. "In the same orderly and scientific spirit", one editorial claimed, "the Arabs evidently intend to train us to identify our interests with their purposes. When we are obedient, the oil will flow. When we are refractory, the oil will stop."[61]

The claim that the United States stood helpless before the power of the Arab producers was also visualized in countless newspaper cartoons published during the embargo winter. Grinning Arab oil sheiks were threatening either frightened US consumers or Uncle Sam with guns shaped like gas-station nozzles. Alternatively, newspaper readers were treated to images of saber-wielding Arabs in traditional garments. Figure 3.1 presents another visualization of US impotence in the face of the oil weapon.

Figure 3.1 Oil pressure, 1974. To many American observers, the 1973 oil embargo seemed to show that the Arab producers could cut off US foreign oil supplies whenever they wanted. Dependency meant that the United States was at risk of losing its political independence in the face of unrestrained producer power.[62]

This editorial cartoon appeared three days after the embargo had been suspended in March 1974. It shows Henry Kissinger in a Middle Eastern desert as he is filling up a large oil truck representing the United States. The hose used by Kissinger runs through the fingers of an oversized, god-like hand belonging to an unknown Arab man who, we are led to believe, could cut off US oil supply with the gentlest squeeze of his fingers. To drive an obvious point home, the reader finds two captions on the cartoon that read "Arab control" and "Embargo lifted conditionally".

Like visualizations of the oil weapon using sabers, daggers or guns, the mighty hand of the Arab man symbolized the unchecked power of the oil producer over its hapless, oil-consuming victim.

Metaphors representing the United States as a puppet, and visualizations of the oil weapon, all served to underline one central lesson learned from the experience of the embargo: dependency meant that the power of the Arab oil producers reigned supreme. As one American commentator put it:

> They have exploited the war to extract nearly $40 billion in additional oil revenues from the rest of the world. The industrialized nations bow down before them, ready to indulge any whim in the desperate hope that in response plentiful oil supplies might be forthcoming.[63]

It is important to note that this reading of the United States as powerless survived the initial shock of the 1973 embargo. Throughout the 1970s, OPEC meetings to discuss production, quotas, and prices were accompanied by a heightened sense of American anxiety and frustration. Consider, for instance, this 1978 editorial in the *Los Angeles Times*, which anticipated an OPEC conference in Abu Dhabi:

> An aura of mystery that attracts speculation and sparks all manner of fears usually accompanies meetings of the oil ministers of the Organization of Petroleum Exporting Countries. The conference to be held beginning Saturday in Abu Dhabi is no exception. To be sure, the OPEC jury of 13 officials will reach the unanimous verdict that oil prices should be raised [. . .] The United States, meanwhile, finds itself in the position of an impotent spectator. It can plead and even scream for moderation, yet, like a fan in a football stadium, it wields minimal influence on the players.[64]

The net effect of these representations in policy and news circles was to construct US dependence on foreign oil as a fundamental threat to US sovereignty and political independence. How could the country continue to formulate foreign policies in its own best interest if any action contradicting the oil producers' political interests could be punished immediately through the use of oil as a weapon? According to this view, the United States was at risk of being subjected to an endless game of political blackmail on the part of the newly powerful producers. The *Washington Post* summed it up concisely: "We cannot be simultaneously dependent upon the Arabs' oil and independent of their political purposes."[65] Jimmy Carter argued similarly that "our national security" rests, among other things, "on the ability of the United States to carry out our foreign policy as a free and independent nation".[66]

In American descriptions of national identity, the notion of independence enjoys the status of one of the nation's founding principles. Beyond the sphere of policy making, dependency also threatened the viability of such American images of Self. Discussing the oil crisis, Zaretsky writes that the new dependency "undermined an important feature of American exceptionalism, namely, the belief that the nation had been founded on independence and that this independence had been sustained throughout the country's history".[67] Betraying this founding principle of the United States over access to energy resources appeared unbearable to many observers. Public debates during the embargo were replete with historical narratives drawing a link between foreign oil and the necessity of affirming independence as a key American value. The American discourse of energy independence sprang from this notion. President Nixon's declaration that energy independence was about reclaiming independence as the "essential nature" of the nation has already been quoted in the previous chapter.[68] Mobil Oil tapped into the same sentiment for its own purposes. On the eve of US Independence Day in 1975, the company published a page-long advertisement in the *New York Times* under the headline "A fourth of July speech in search of an orator." The advertisement contained a fictional Independence Day speech which linked the origins of the nation to its present condition of dependency: "Back in 1776, our forefathers brought forth on this continent a new nation—not only politically free but 100 percent independent of any foreign power for energy supply." The advert then argued that the founding fathers would be outraged to learn that the country risked its future prosperity and independence because of political failures to develop new domestic energy resources. "So join with me, fellow citizens, on this great anniversary of our independence, in helping to rededicate America to regaining energy independence—that precious asset our forefathers enjoyed."[69] A similar point was made by former

Secretary of the Interior Walter Hickel. "Independence was America's birthright," he wrote in a 1974 article. "It's up to us that we don't sell that birthright for a barrel of oil."[70]

Foreign Oil and American Leadership

A third threat associated with foreign oil dependence is encapsulated in the claim that dependency diminished the power and international leadership of the United States. In some ways, this argument is closely linked to what was discussed in the previous section. If the direction of US foreign policy can be prescribed to some extent by foreign oil producers, this cannot be without adverse effects on the international power position of the United States. But dependency also impacted on the capabilities of the country beyond its power to project military force abroad or to defend itself. Foreign oil threatened American leadership in the service of the international community at large. Within the foreign policy community, dependency was described as a threat to America's identity as a Cold War superpower, as the leader of the free world and, implicitly, as the indispensable nation whose power to shape the world is unrestrained by the wider world itself.

One important location of this debate about foreign oil and power was the academic community. In classical realist and liberal accounts of international relations, the power of the State was usually defined in terms of material capabilities. Population, economic wealth and, in particular, military capabilities were seen as key measures in this regard. In this calculus of power, oil entered the equation above all as a fuel for the internal combustion engines central to the modern military machinery. From this perspective, access to sufficient quantities of oil had to be seen as a precondition for the ability of any great power to project its military force abroad. In addition, as a driver of industrialization and economic growth, oil indirectly contributed to state power and the potential for great power hegemony.[71] It is no coincidence, David Painter argues, that both superpowers emerging from the upheaval of World War II were self-sufficient in oil.[72]

The oil crisis complicated this picture, however. Many scholars noted that the links between oil and power were more complex. Despite their utter insignificance as military and industrial powers, it seemed that the oil producers had somehow transformed oil into a power resource that could even bring a giant like the United States to its knees. This left a deep impression on the United States, both in the academy and the world of politics. As Yergin later wrote, "The very substance of power in international politics seemed to have been transmuted by its oleaginous reaction with petroleum."[73] Such momentous changes required explanation. They triggered a debate about the notion of power itself. Hans Morgenthau, the leading realist author of the day, argued that OPEC's control of oil "threatens to overturn centuries-old patterns of international politics." The functional relationships between political, economic and military power, he further argued, "have been disturbed—one might even say destroyed—by the recent use of oil as a political weapon."[74] This debate led to the formulation of alternative approaches to understanding power. Yet, the mere fact that it took place reflected a widespread concern that, as part of the energy crisis, power had somehow shifted away from the United States.

These concerns were also voiced by policy makers, journalists and other practitioners of international affairs. They offered a narrative about the decline of American power that was saturated with references to the cultural foundations of the hydrocarbon society. In particular, they argued that oil had made the United States a great nation. The time-honored notion of abundance sat at the heart of this narrative. As the first major producer of petroleum, the United States was said to look back on a history of oil wealth that had been a major source of strength throughout the twentieth century. As the *Chicago Tribune* noted in 1973, self-sufficiency in energy had been

> one of the greatest advantages this country has had over the years. We have never been utterly dependent on imports for any of the raw materials essential to our way of life. In our relations with foreign countries, therefore, we could act on the basis of reason, not of dire necessity.[75]

According to this historiography of oil and power, oil had made the United States strong. But postwar changes in oil production patterns had then affected the balance of power between producers and consumers. US domestic oil production dwindled relative to oil demand, and with it America's source of strength. One author even argued that "From 1947 to 1973, the shift of power is exponential."[76] Slowly but surely, the United States had moved into a position of dependency. It was a position in which oil functioned no longer as a source of strength but as a source of weakness.[77] As the *Washington Post* argued during the 1979 energy crisis, "If the Iranian chaos should be teaching us anything, it is the degree of our vulnerability to oil imports. We are in danger of becoming a second-class power, beholden to the sheiks in Riyadh."[78]

This narrative was closely connected to the view that US dependence on foreign oil endangered the Cold War identity of the United States as the leader of the free world. As the following chapter will explain in some detail, the energy crisis appeared to many observers in the United States as a serious threat to the international order that the United States had created and dominated after 1945. In a speech at the Washington Energy Conference in 1974, Henry Kissinger called the energy crisis an "unprecedented challenge to our prosperity, and to the entire structure of international cooperation so laboriously constructed over the last generation", adding that it posed "fundamental questions about the future of the developing countries, the prospects for economic growth of all nations, and the hopes for global stability". Only international co-operation could avert a collapse of world order similar to what had happened in the 1930s. The "ultimate challenge" raised by the crisis, according to Kissinger, "is to the fragile fabric of international principles and institutions".[79] American dependence on foreign oil was only one element in this picture, but an important one. Dependency seemed to overturn established relationships between the core countries and the periphery of the Western international order. Although the United States was less dependent on foreign oil than most other Western democracies and Japan, as the leader of the Western world it had special responsibilities. Could the United States remain dependent on foreign oil and continue protecting and promoting the vital interests of the Western world? Doubts about this were expressed in the highest echelons of the US Government. Only weeks after the beginning of the embargo, President Nixon said this in an address to the nation:

> Today the challenge is to regain the strength that we had earlier in this century, the strength of self-sufficiency. Our ability to meet our own energy needs is directly limited to our continued ability to act decisively and independently at home and abroad in the service of peace, not only for America but for all nations in the world.[80]

In 1976, Vice-president Rockefeller declared that another embargo would seriously impair "our capacity to meet our responsibilities to the other nations of the free world". Rockefeller asserted:

> The continued dependence upon foreign sources of oil could cause us to lose credibility with our allies. They would be justified in asking whether or not we would support their interests against those of our oil suppliers. Our continuing dependence on imported oil threatens our leadership in the free world.[81]

Nixon and Rockefeller portrayed dependency as a threat to the status of the United States as a great power. The danger was not so much that dependency endangered the superior military force that the US Government could project abroad. On the contrary, it was *despite* America's unquestioned military superiority that power had shifted away, and it was the helplessness of the military giant vis-à-vis the weak but rebellious oil producers that was portrayed as causing the embarrassment. This is perhaps best evident in the repeated claims that avoiding dangerous dependence on foreign oil was vital to the country's "self-respect" and "dignity".[82] The most explicit articulation of this threat was not provided by a president of the 1970s but by George H. W. Bush in the early 1990s. As Bush explained,

> Our ability to function effectively as a great power abroad depends on how we conduct ourselves at home. Our economy, our Armed Forces, our energy dependence, and our cohesion all determine whether we can help our friends and stand up to our foes. For America to lead, America must remain strong and vital.[83]

Foreign Oil and Threat after the 1970s

At first sight, very little seems to have changed in American descriptions of the dangers of foreign oil since the 1970s. Although the threats linked with foreign oil dependence are almost never stated as explicitly today as they were in the wake of the energy crisis, we find in recent statements of policy makers, journalists and other speakers traces of almost all the major quasi-causal arguments described above. But beneath the surface, a few important changes have occurred. The remainder of this chapter will present both these continuities and discontinuities.

As in the 1970s, today's descriptions of foreign oil dependence are situated in broader political contexts. In the 1970s, the Cold War as well as the debate about American decline provided important frames of reference for the definition of dependency. In the early twenty-first century, the continued globalization of energy markets, the rise of new powers such as China, India and Brazil, worries about climate change, and the beginning US boom in the development of unconventional sources of oil and gas provided important contexts for the way in which energy was debated in Washington.

At the turn of the century, rising prices brought energy back onto the political agenda. The two terms of President George W. Bush coincided with what may be called the "long" oil price shock of the 2000s. Average oil prices rose from $30 per barrel in January 2001 to $65 per barrel in January 2006 and, finally, to a peak of $133 per barrel in July 2008. Upward pressure on the price of oil resulted from the combined effects of increasing global oil consumption, most notably in emerging economies; the volatile situation in the Middle East after the 2003 invasion of Iraq; and from what Yergin calls "aggregate disruption"— supply losses caused by strikes in Venezuela, violence in the Niger Delta and Hurricane Katrina in the United States.[84] While concern with rising energy prices was already evident toward the end of the Clinton presidency, it was particularly after 2004 that energy moved to the top of the political agenda.

Another important development occurred in the wake of the terrorist attacks of 11 September 2001. The focus of US foreign policy concerns shifted decisively toward the Middle East. Political Islam and fundamentalist terrorism had been a policy concern in the United States since the 1980s. But the attacks on New York and Washington sent shock waves through the political fabric of the nation, including a shock to oil's discursive economy in the United States. This had significant effects on the way in which the country's oil-related relationships with the producers in the Middle East were discussed in Washington. There was a pervasive sentiment among US policy makers that the attacks had "changed everything", including the dangers of foreign oil and the appropriate responses to this long-standing challenge.[85] Linking US dependence on foreign oil to the threat of terrorism was the most important new articulation of danger in this debate. Both rising oil prices and the emergence of international terror as the major security issue in US politics occurred during the administration of President George W. Bush. This is why no other US president since the 1970s talked as explicitly and as often about the dangers of foreign oil as Bush. Barack Obama's presidency, in turn, was shaped more by the concurrent boom in unconventional sources of oil and gas. In the context of rising production quantities, falling prices and a lessened dependency, the dangers of foreign oil have appeared much less urgent and dramatic.

A Continuity of Threats

On the surface, the continuities between the 1970s and the present-day discourse about the dangers of foreign oil dependence appear striking. When a deputy secretary of energy claims before a Senate committee that foreign oil dependence means that "we are at the mercy of events and decisions over which we have often limited—sometimes no—control", this statement should be regarded as an essentially timeless feature of US foreign oil debates since the 1970s. This is true both regarding the content of the statement—the claim that the United States is helpless and vulnerable in the face of dependency—and regarding the particular choice of words—to be "at the mercy" of outside forces.[86] The critical importance of the 1970s for defining the way in which dependency is currently being discussed is further underlined by the omnipresence of the 1973 crisis in contemporary statements about energy. Policy makers and experts like to evoke the "lessons of the energy crisis", and even those who argue that much has changed tend to discuss these changes against the background of the 1970s.[87]

The most important continuity in debates about foreign oil is the fear that foreign oil supply may suddenly be interrupted. Like his predecessors, President George W. Bush frequently referred to the possibility of supply interruptions as a major reason why dependence on foreign energy was a matter of national security. The consequences of such interruptions were represented in terms strikingly similar to those found in the 1970s.

Consider, firstly, foreign oil dependence as a threat to the US economy and the American way of life. Policy debates about oil after the 1970s still tended to start from the premise that energy was the essential commodity underpinning the US economy. Another 1970s-type oil shock, the argument went, "would have profound effects that could severely threaten the foundation of America's economic prosperity".[88] Inherent in such anticipated economic upheavals was a potentially existential threat to American society as a hydrocarbon society. The threat was not that American society could be physically extinguished but that a prolonged interruption of supply would violently change it beyond recognition. As oil expert Michael Klare asserted in his 2004 book about dependency, the key source of danger arising from the nation's foreign oil dependence lies in "the centrality of cheap and abundant petroleum to the vigor and growth of the American economy and to the preservation of a distinctly *American* way of life". Without the continued supply of petroleum, above all as gasoline for the privately owned car, at prices that middle-class Americans can afford, this American way of life "could hardly survive".[89]

These oil experts painted a picture in which the continuing American dependence on foreign oil appeared as a direct threat to consumer choices and lifestyles that characterize the American way of life. In the following extract from a *Washington Post* editorial, which appeared in the direct aftermath of the 9/11 attacks, journalist Robert Samuelson described oil as essential to this way of life. In his view, the terrorist attacks were yet another warning that the country's reliance on unreliable foreign supply threatened American society at large:

> The American way of life doesn't depend on $1 or $1.50 gasoline. It does depend on reliable sources of energy [. . .] The dangers have been obvious for years, and our failure to react ought to be a source of deep national embarrassment. This is a long-term problem; anything we do now won't have significant effects for years. But if we fail to heed the latest warning, the neglect would be almost criminal.[90]

Besides oil experts and journalists, US policy makers also suggested that foreign oil dependence was a threat to the US economy so severe that it endangers the American promise of individual opportunity and social mobility. In his 2007 State of the Union Address, President George W. Bush said that "extending hope and opportunity depends on a stable supply of energy that keeps America's economy running and America's environment clean". Explicitly, the continued reliance on imported oil was depicted as a threat to the future economic prosperity of the country:

> For too long, our nation has been dependent on foreign oil. And this dependence leaves us more vulnerable to hostile regimes and to terrorists who could cause huge disruptions of oil shipments and raise the price of oil and do great harm to our economy.[91]

Secondly, there has been continuity with regard to the representation of foreign oil dependence as a threat to the independence and sovereignty of the United States as reflected in the country's ability to formulate a foreign policy of its own choosing. Because US oil import requirements always need to be taken into account when formulating foreign policies, the Government has been constrained in its relations with the major producers. Former President Jimmy Carter pointed out on the occasion of a Senate hearing in 2009 that, as regards foreign oil dependence, the basic "strategic issues [. . .] have changed very little during the past three decades". As Carter put it, "There is little doubt that, in general terms, we are constrained not to alienate our major oil suppliers."[92]

The assumption that dependency caused vulnerabilities and constraints in US foreign policy has been a continuous theme in official representations of the issue. In the 2000s, President Bush frequently pointed out that "it affects international politics, to a certain extent, to be dependent on oil".[93] On another occasion, Bush described this threat more fully, yet slightly less eloquently:

> If you're dependent on oil overseas, it means that [. . .] you become vulnerable to the activity of a hostile regime. In other words, somebody doesn't like us; they produce the oil; they decide to do something about it; they can affect us. That's—when I talk about the national security risks, that's what I mean.[94]

America's vulnerability to the use of oil as an instrument of politics, as Bush described it, meant that producers could use oil to blackmail the United States. Not only could US actions abroad be punished by oil-producing states, the United States was also constrained to avoid policies that threatened the smooth operation of the international oil markets.

While the assumption of policy constraints resulting from dependency is well established, professional observers of US foreign policy have represented these constraints in increasingly dramatic terms in the early twenty-first century. In 2005, the *Wall Street Journal* identified an "omnipresence of oil in America's foreign calculations" and observed, "If it seems that the odor of oil hangs over America's whole international agenda right now—well, that's about right."[95] Recent analyses of US foreign policy are replete with such representations of the disproportionate and harmful influence of petroleum on American activities abroad. A particularly forceful instance of this theme appeared shortly after the events of 11 September 2001 in a *Washington Post* editorial, whose author deplored the way in which

> our foreign policy is captive to oil. If you were to draw a flow chart tracing every terrorist attack against the United States back through every foreign alliance, military mutual aid pact, joint military exercise, bloody political coup, intrigue and tacit alliance, the lines would generally flow back to one common factor. Oil.[96]

This view represented the extreme end of the well-established assumption that dependency constrained US foreign policy. The need to secure the global flow of oil has become the paramount foreign policy objective of the United States, even where this is at odds with other important national interests. Unlike in the 1970s, more recent energy debates have frequently contained this stark assumption. In addition to constraining the formulation of foreign policies, the argument goes, the nation incurs significant costs in blood and treasure as a result of its unabated foreign oil dependence. A straightforward articulation of this link between dependency and foreign policy is contained in the following statement by civil-society activist Tom Collina during a congressional hearing in 2005:

> The greater our dependence on oil, the greater the pressure to protect and control that oil. The growing American dependence on imported oil is the primary driver of US foreign and military policy today, particularly in the Middle East.[97]

The principal burden for this task falls on the US military. Several studies have attempted to quantify the dollar costs of protecting the global oil transportation infrastructure with military means, and have given estimates ranging from $13 billion to $143 billion per year.[98]

These policy experts have frequently portrayed US foreign oil dependence as a key item in the nation's military budget and, more importantly, as an important cause of US military action abroad. In 2004, Michael Klare, for example, asserted that foreign oil dependence was behind what he saw as the increasing militarization of US policy in the Middle East. "Slowly but surely", Klare observed, "the US military is being converted into a global oil-protection service."[99]

This representation of dependency can be understood as another threat to prevalent descriptions of American identity. As Chapter 5 will explain more fully, an overwhelming majority of Americans have seen the use of military force to secure the flow of oil to be a quintessentially un-American foreign policy. To the extent that dependence on oil is understood to drive such a policy, a gap opens up between America's self-representation as a nation with a deep sense of mission and morality and the observable reality of US actions on the ground. This, in essence, is behind statements arguing that dependency is "compromising American values and principles".[100] Again, Collina made this point explicitly:

> Now imagine America free from the burden of protecting our stake in Middle East oil, allowing us to reduce our military footprint in the region and our exposure to terrorism. We could then base our foreign policy on the ideals that make this a great nation, like global peace and security, freedom and democracy.[101]

Similarly, recent predictions that the United States may achieve energy independence as a result of the American energy revolution have been celebrated because of the political implications that self-sufficiency may yield. The United States, argues one group of energy analysts, would finally be "freed from the shackles involved in sacrificing a values-driven policy focusing on human rights and democratic institutions in order to secure cooperation from resource-rich despotic regimes".[102] For the time being, however, the question remains: how can the nation's dependence on foreign oil be reconciled with the idea, flowing directly from the discourse of American exceptionalism, that the United States is a force for good in the world? Of course, the idea that foreign policy could be based entirely on moral convictions instead of hard national interests is a fiction. Nevertheless,

US dependence on foreign oil threatens the viability of this narrative and thus challenges a well-established discourse about the moral foundations of US foreign policy.

A New Logic of Threat

The previous chapter showed that American debates about foreign oil in the twenty-first century are shaped by a certain ambiguity about the best definition of dependency. The established view since the 1970s defined the problem of dependency as foreign oil dependence. Since the 1990s, however, this definition has been challenged by an emerging emphasis on the problems associated with all forms of oil consumption. In effect, the problem of dependency came to be seen in a much broader sense as US dependence on oil in general. Today, it is often unclear whether individual statements about danger refer to foreign oil dependence specifically or to all forms of oil dependence.

But, in fact, matters are even more complicated today. Even where claims about national security threats are directed at foreign oil specifically, a clear distinction between the two definitions of dependency increasingly breaks down. The reason for this additional layer of ambiguity is that foreign oil is associated with threats in new ways today. In the 1970s, the link between foreign oil and threats to national security was constituted by two elements. Firstly, not all foreign oil was said to be dangerous but only that oil that the United States actually and physically imported. This seemed so self-evidently true that no one ever bothered to spell it out. Yet, the logic behind this assumption was reflected in the standard policy proposals to deal with US foreign oil dependence: diversification. Greater security could be achieved by shifting imports from dangerous OPEC sources to secure sources like Canada. In the 1970s, such an approach to mitigating the dangers of foreign oil was suggested, for example, by Nixon's task force on oil imports.[103] In the twenty-first century, this is still a widely held interpretation of dependency. For instance, the George W. Bush Administration argued along similar lines in its influential 2001 *National Energy Report*. The energy policy set out in the document specifically aimed at the "diversification of energy supplies" to enhance US national security, focusing on such areas as the Western hemisphere but also west Africa and the Caspian Basin.[104]

The second element linking oil imports to threats to national security consists of what Michael Klare has described as a logic of "more dependence, more peril". According to this idea, the dangers inherent in foreign oil dependence grow with increasing quantities of imported oil.[105] Reflecting this assumption, the share of imported oil in total US oil consumption has come to function as the principal measure of the foreign oil threat. Observing the movements of this variable became the standard way of assessing whether the country was on the right or wrong track as far as the problem of dependency was concerned. The following statement by President Ford is a typical illustration of the logic of more dependence, more peril:

> The United States is dependent on foreign sources for about 37 percent of its present petroleum needs. In 10 years, if we do nothing, we will be importing more than half our oil at prices fixed by others—if they choose to sell to us at all. In $2\frac{1}{2}$ years, we will be twice as vulnerable to a foreign oil embargo as we were two winters ago.[106]

As a quantification of the foreign oil threat, this variable is overly simplistic because it ignores other important elements like spare production capacity, strategic stockpiles, private inventories and domestic refinery capacities. Yet, this has done nothing to impede the widespread use of this figure as a symbol of the threats of dependency. The US Department of Energy, for example, maintains a regularly updated website that keeps track of the nation's exposure to the dangers of foreign oil. The website is called "How dependent are we on foreign oil?" With its focus on the share of net imports in total US petroleum, this document is a great example of the logic of more dependence, more peril.[107]

More recently, however, the twin assumptions of threat through physical importation and more dependence, more peril have been challenged by an alternative view of dependency. This alternative reading is structured around the notion that oil is a *fungible* commodity—that is, that oil from one source can usually be replaced by oil from another source because of its similar properties. As a result of oil's fungibility and the global nature of energy flows, energy professionals argue, developments anywhere ultimately filter through the world oil market in its entirety. The emergence of the logic of fungibility has had important consequences for the articulation of links between foreign oil and danger. Most importantly, as one study puts it, "Because oil is fungible, the geographic origin of US imports is irrelevant."[108] The United States, by implication, is exposed to the dangers of foreign oil not because it imports oil from certain unsafe sources, nor even because it imports any oil at all. The nation is at risk simply by virtue of the fact that it is fully integrated in the global oil market. This undermines the notion that the link between foreign oil and danger is constituted by actual importation as well as the logic of more dependence, more peril. According to the logic of fungibility, foreign oil is dangerous because the United States consumes oil at all and, thus, participates in the global oil market. "A country's exposure to world price shocks", as two observers explain, "is thus a function of the amount of oil it consumes and is not significantly affected by the ratio of domestic to imported petroleum."[109] In effect, as far as the major threats to US national security are concerned, the diversification of US supplies and a focus on reliable sources is not in itself a comprehensive solution. As another group of experts argued, "The United States could shift all its purchases to sources that are relatively safe politically, such as Canada and Mexico, and it would still not be protected. The global price is what matters most."[110] Ultimately, the logic of fungibility presents a challenge to the notion of foreign oil itself as, increasingly, the viability of a meaningful distinction between oil produced at home and oil produced abroad breaks down.

These arguments are now commonplace. In the 1970s, oil economists like Morris Adelman had argued on similar terms that "even zero imports" would leave the United States vulnerable.[111] Four decades ago, however, such views could not challenge the twin assumptions that threats resulted from physical importation and that the logic of more dependence, more peril applied. Today, by contrast, the logic of fungibility is gradually replacing the established logic of the foreign oil threat. The result is even greater ambiguity about the precise meaning of statements about foreign oil and danger.

These developments have begun to leave their mark on established understandings of energy security. In the 1970s, based on the experience of the first and second oil shocks, the risk of interruptions of the nation's foreign oil supply had been predominantly attributed to politically motivated cut-offs or accidental interruptions in the countries from which the United States imported the lion's share of its oil. In the early twenty-first century, by contrast, anxiety about possible interruptions is no longer limited to fears about these countries. Neither is it limited to embargoes or political cut-offs. All manner of reasons now feature prominently in statements about the risks of supply interruptions, including the sabotage of pipelines, the closure of sea lanes, weather phenomena, interstate war in oil-producing regions, civil strife or strikes in oil-producing regions, and accidents at production or transportation facilities. Observing this expansion of vulnerability, Yergin writes that ensuring energy security has become "an awesome task". He explains:

> The current model of energy security, which was born of the 1973 crisis, focuses primarily on how to handle any disruption of oil supplies from producing countries. Today, the concept of energy security needs to be expanded to include the protection of the entire energy supply chain and infrastructure.[112]

Hence, while the basic underlying fear of supply interruptions has remained unchanged, this threat tends to be articulated much more broadly today. The logic of fungibility implies that interruptions anywhere in the world threaten the energy security of the United States. Given that there are an almost infinite number of potential vulnerabilities in the global energy infrastructure, the security of supply appears potentially much more precarious than in the 1970s. Felix Ciută observes a similar trend, arguing that there is an emerging "total" logic of energy security according to which "energy security means the *security of everything*".[113]

The New Threat of Terrorism

In the wake of the Al-Qaeda attacks on the World Trade Center in New York and the Pentagon in September 2001, Islamic terrorism instantly became America's top national security problem. Beyond the question of how the United States should react to the first foreign attack on its soil since Pearl Harbor, the events sparked off a broad and very critical re-evaluation of the country's relationship with Islam and the Arab countries of the wider Middle East in particular. Since oil was a key issue that linked the Arab world and the United States, oil imports naturally became an important facet of this debate. More specifically, the oil resources of the Middle East became strongly associated with radical Islam and violent opposition to the West. Oil and terrorism had sometimes been connected in US foreign policy debates in the 1980s and

1990s, but in the wake of the 2001 attacks the claim that Arab oil and Arab terror were intimately linked became central to the discourse about the dangers of foreign oil.[114]

This argument came in different forms. To begin with, the logic of fungibility implied that in order to hurt the United States the terrorists could cause a disruption of the flow of oil by attacking any one of the many important nodes in the global energy supply chain. Terrorism became one more cause for a potentially devastating interruption of the country's oil supply. As Goldman Sachs Vice-chairman Robert Hormats explained before the Senate in 2007, "The American economy remains highly vulnerable to supply disruptions in oil exporting nations; these could result from acts of terrorism, political instability, efforts to use oil as leverage, or natural calamities".[115] In parts of the US foreign policy community, Islamic terrorists were described as powerful actors by virtue of their ability to use oil as a weapon to attack the United States. Unlike in the 1970s, however, the mechanism through which oil became a weapon was not via direct control over the production, price and sale of oil but rather through the reliance of the United States on a far-flung and inherently vulnerable global energy supply chain. "America's enemies know the oil is the Achilles heel of the developed world and that the oil transportation system is vulnerable," argued a report issued by the National Resources Defense Council in 2005.[116] All manner of scenarios involving pipelines, tankers or production facilities have since been anticipated, in which terrorists may use American dependency to inflict damage on the country. What, for example, if terrorists set off a "dirty bomb" in Ras Tanura, the most important Saudi Arabian oil port?[117]

Such concerns also seeped into official representations of dependency. George W. Bush worried about the potential consequences of a terrorist attack on the global energy infrastructure. On one occasion, he explained:

> Dependence on oil, as well, means that if a terrorist were able to destroy infrastructure somewhere else in the world, it's going to affect what you pay for at the gasoline pump. In other words, as we learned, the terrorists attacked us in brutal ways; they attacked us by flying airplanes into our buildings. But we're also vulnerable to attacks through infrastructure around the world that could cause your price, the price you pay, to go up.[118]

The continued US dependence on foreign oil has been linked to terrorism in an even more fundamental way, however. The country's need to import oil, it is sometimes argued, transfers a significant amount of money from the United States to the Middle East, some share of which inevitably ends up in the coffers of Al-Qaeda, Hezbollah or Hamas—or in government accounts used to spread extreme forms of religious ideology, such as Wahhabism. Through its dependence on foreign oil, the United States has in effect become a sponsor of terrorism of sorts. America is funding both sides in the war on terror. In this regard, Senator Conrad Burns coined the term "rogue oil" in a *Wall Street Journal* article, by which he meant that portion of foreign oil that is associated with the undemocratic regimes of the Middle East and perhaps also Venezuela. These nations were characterized by their hostility to the United States and close connections to anti-Western extremism and terrorism. As the senator explained, "The truth is that oil and terrorism are as much entangled as radical Islam and terror. If bastardized religion is terrorism's inspiration, oil revenue is its succor."[119]

Another fine example of this claim is contained in the following assessment by Gal Luft, a well-known American energy expert, in the context of sharply rising oil prices in 2005. Before a Senate committee, Luft argued that US imports of oil not only constituted a major burden on the US trade balance and a tax on the American people, it also helped "strengthen the jihadists". As he explained:

> In November 2001, a barrel of oil was selling for $18. In less than 4 years the price jumped to $70. This means that Saudi Arabia [...] receives an extra $\$\frac{1}{2}$ billion every day from consuming nations and Iran [...] an extra $125 million. This windfall benefits the nondemocratic governments of the Middle East and other producers and finds its way to the jihadists committed to America's destruction as petrodollars trickle their way through charities and government handouts to madrassas and mosques, as well as outright support of terrorist groups.

Hence, while the US government was using taxpayers' money to support the defense of the country against its enemies, at the same time the country was funding those enemies through oil revenues. Luft concluded:

> 4 years after September 11 it is essential that we view our geopolitical situation in the context of our oil dependence and realize that it will be extremely difficult to win the war on terror and spread democracy around the world as long as we continue to send petrodollars to those who do not share our vision and values.[120]

In the 1970s, the dangers of foreign oil arose mainly from the rebellious and unreliable nature of the OPEC suppliers as well as the general political instability of the Middle East. Today, the foundation of the Middle East's dangerousness has shifted. While concerns over the power of the oil cartel have diminished, terrorism and, more generally, radical hostility to the United States have appeared as major sources of threat. Today, these dangers appear greatly amplified because the Middle East itself has become, in the eyes of many observers, a more dangerous place, populated by what President Bush described as "unfriendly regimes",[121] "rogue states",[122] "hostile regimes and [...] terrorists".[123] The following chapter will expand on this argument.

Conclusion

This chapter presents some of the most commonly suggested links between US foreign oil dependence and threats to US national security as they were articulated in the wake of the 1973 energy crisis. In this view, dependency threatened the viability of the American way of life as enshrined in the material patterns of energy consumption of the American middle-class family, as well as the discourses that have given meaning to these practices. Dependency also threatened the founding principles of the American national project, subsumed under the umbrella of American exceptionalism: the independence and sovereignty of the nation in its foreign affairs and America's indispensable role as the leader of the free world. Finally, and more recently, dependency and threat have been linked via the issue of international terrorism.

Charlotte Epstein observes that the discursive construction of meaning is a messy, bloody and dynamic process.[124] Presenting only the major results of these struggles always involves the risk of conveying a much neater, overly polished picture of how conflicting and ambiguous claims about causes and effects come into being, evolve, clash, disintegrate or fade away over time. The case of the quasi-causal arguments about foreign oil and threat are no exception. Many nuances and subtleties, and additional themes as well as dissenting voices have been omitted from the account in this chapter. Each and every claim presented above has been contested and rejected by someone at some point. This includes even the most foundational quasi-causal links between oil and threat. For example, doubt is often cast today upon the idea that the 1973 oil crisis was the sole or even primary cause of the economic downturn that ended the phase of spectacular postwar economic growth in the United States and other Western countries. According to Fiona Venn, for example, 1973 was a rather "symbolic" event marking the end of the old Bretton Woods system. Problems in the world economy had begun to appear much earlier.[125] Other energy professionals argued that the historical evidence suggests that targeted oil cut-offs are not the effective political weapon they were made out to be in the 1970s. "Oil embargoes", the authors of a RAND study conclude, "have been an ineffective tool for advancing foreign policy goals."[126] Some international relations scholars, among them Kenneth Waltz, have pointed out that the pessimistic reading of the 1973 energy crisis was unfounded because, as it turned out, the American power position was not significantly affected by the country's foreign oil dependence.[127] Most recently, claims about dependency and terrorism have been questioned. It is sometimes said that there is no hard evidence corroborating the existence of a link between oil revenues and terrorism in producing countries. It is further claimed that terrorist attacks are usually low-cost enterprises that rely on personnel and know-how much more than on significant funds.[128] As the introduction to this volume shows, skepticism about the accuracy of arguments linking dependence and threat has led some energy professionals and policy makers to diagnose the United States with a severe "obsession" with foreign oil dependence.

These objections have not been presented in detail here, nor has it been the aim of this chapter to assess the factual accuracy of common claims about the dangers of foreign oil. Dependency is a discourse whose power has not relied on the empirical validity of its central truth claims but on its intricate relationship with US constructions of national identity. In order to understand the American "obsession" with the dangers of foreign oil, this chapter suggests, it is necessary to examine the nature and origins of the threats said to result from oil imports. These claims emerged during the energy crisis of the 1970s in response to events which constituted a shock to the United States not only in the economic sense but also in a cultural and discursive sense. In an environment in which anxiety and uncertainty appeared pervasive, the notion that the country's dependence on Middle Eastern oil constituted an essential threat to the security of established lifestyles and descriptions of Self moved into the heart of US foreign policy debates. Irrespective of whether this is justified by sober analysis of the situation, echoes of these claims still shape American views of dependency today. The persistence of the discourse of dependency is testament to the power inherent in this particular reading of the energy crisis, thus corroborating the argument that there is no necessary correlation between the objective existence of a threat and its constitution as such in foreign policy discourse.

CHAPTER 4

THE FOREIGNNESS OF FOREIGN OIL

Not all foreign oil is equally dangerous. So far, the discussion has focused on American views of US foreign oil dependence as a dangerous and threatening condition. The problem of dependency is more complex and nuanced than that, however. The notion that not all foreign oil is equally dangerous provides an excellent starting point for reaching a deeper understanding of the American cultural discourse of dependency. This chapter will examine just what it is that makes some foreign oil so dangerous.

The assumption that some sources of oil are more secure than others has long been a staple in American debates about oil. In the introduction to his book *Blood and Oil*, Michael Klare discusses what he sees as the two most worrisome trends concerning the country's foreign oil dependence. The first issue Klare identifies is America's seemingly limitless thirst for energy; the second and more important problem, as he sees it, lies in the fact that

> more and more of our imported oil will be coming to us from countries that are unstable, unfriendly, or located in the middle of dangerous areas (or some combination of all three). And it is *this* trend, more than any other, that makes our reliance on foreign energy so worrisome. After all, we wouldn't be so anxious about our growing dependency if all our energy was coming to us from Canada, Norway, Australia, and other friendly producers located in peaceful parts of the world.[1]

In much the same way, the US Government has long based its evaluation of the country's foreign oil dependence on a sometimes tacit, sometimes explicit distinction between secure and insecure sources of imported oil. In 1970, Nixon's task force report on the Mandatory Oil Import Program argued that preferential treatment for Canadian, Mexican and Latin American oil would be appropriate given that these were "the most secure" sources of foreign oil.[2] Four years later, the *Project Independence Report* noted that the degree of US energy vulnerability was a function of "the level of insecure imports" and then proceeded to quantify and compare secure as against insecure supply.[3] After the twin oil shocks of the 1970s, it became received wisdom that OPEC countries, and the Persian Gulf suppliers in particular, constituted the single most insecure source of US imported oil, whereas deliveries from Canada, Mexico or Venezuela were considered to be relatively secure. In 1982, Melvin Conant, an influential energy expert, stated that "the American interest is clearly in maximizing supply from these [the latter] countries to reduce dependence on the Gulf".[4] For decades, the formulation of federal energy policy has been guided by this view. In 1986, Ronald Reagan was pleased to announce that the amount of US imports had dropped in absolute terms. Yet, "even more significant", according to the President, was that "we've changed who we buy our imported oil from". Because more and more oil was coming from the Western hemisphere or North Sea fields instead of the Middle East, Reagan explained, "our supplies won't be as vulnerable to international politics as they've been in the past".[5] In the 1990s, Bill Clinton made the diversification of oil supply part of the National Security Strategy.[6] Under George W. Bush, diversification outside the Persian Gulf remained a central objective of energy policy as formulated in the 2001 *National Energy Report*. While the goal of diversification included oil from virtually all available sources around the globe, secure or insecure, at the heart of this strategy rested the goal of "closer energy integration" with secure and reliable suppliers like Canada and Mexico.[7] As a result of this broad, continuous policy approach, non-OPEC oil became an increasingly important element in US petroleum imports, and Canada became the single most important source of imported oil. In 2013, Canada alone provided about 32 percent of the oil that the United States was importing.[8]

The assumption that the dangerousness of foreign oil is linked to its geographical origin has been central to the distinction between secure and insecure sources of imported oil, and has served to justify preferential treatment for imports from some countries but not others. Geography, national borders and spheres of sovereignty thus assign meaning to oil in two stages: US national borders first define oil as "domestic" or "foreign". All foreign oil is then again divided into Canadian oil, Saudi Arabian oil, Norwegian oil, and so on.

If oil's dangerousness is linked to individual producing states in this manner, an interesting series of questions follows: What determines the security or insecurity of any one source of oil? What is it that makes Canadian oil so much safer than Saudi Arabian oil? Although the question is rarely, if ever, formulated in this manner by policy makers and experts, we can extract a number of factors determining oil's dangerousness from their statements: oil from some countries comes a long way across oceans, often through narrow and vulnerable choke points like the Strait of Hormuz; some oil-producing countries have proven unreliable in the past; others have troubled or even hostile political relationships with Washington; yet others are internally unstable, threatened by other states, or located in volatile regions of the world; some countries fall into all of the above categories.

Without casting doubt on the general usefulness of such risk assessments, the following pages will show that there is something else at play as well. Or, to be more precise, that such assessments of risk are made *as part* of a larger, subtler and much more fundamental process of making sense of the world. The question of oil's dangerousness is intimately connected to constructions of American identity.

This argument rests on a conceptualization of national identity that emphasizes the performative and relational nature of identity.[9] To say that identity is constituted performatively is to say that it is not fixed, stable, simply given or inherent in the object to which it belongs. National identity has to be constructed and constantly remade and affirmed in order for it to persist. As Benedict Anderson so famously observed, nations are "imagined communities".[10] To say that identity is relational means that it is established by drawing a boundary between Self and Other. Identity is created in relation to difference, to what it is *not*. This means that the construction of American national identity necessarily involves making claims about what the rest of the world is like—the peoples, places and cultures in it. Claims about insecurities, dangers and threats play an important role in this process because they enable particularly powerful distinctions between Self and Other. William Connolly argues that establishing an identity involves transforming innocent difference into "otherness, into evil, or one of its numerous surrogates. Identity requires difference in order to be, and it converts difference into otherness in order to secure its own self-certainty."[11] From this perspective, foreign oil is not simply that oil that has to cross the geographical borders of the United States but rather that oil that has to cross the culturally constructed boundary between Self and Other. And these geographical and cultural boundaries are by no means identical.

The central argument here is this: foreign oil is dangerous because it comes from places that have themselves been constructed as foreign and dangerous as part of an ongoing process of establishing American national identity. Foreign oil comes from outside the culturally constructed boundary separating the American Self from its various cultural Others. In order to move beyond a simple material, geographical understanding of what "foreign oil" is, and in order to emphasize how the dangers of imported oil are connected to constructions of American identity, it is perhaps best to rephrase the statement made at the outset of this chapter. Instead of saying "not all foreign oil is equally dangerous", we should put it as follows: all foreign oil is dangerous, but not all imported oil is equally foreign. This is not just a nice turn of phrase. As we will see, oil's culturally constructed foreignness has helped to constitute the meaning of US dependence on foreign oil as well as the policies formulated to deal with this problem.

This, then, is the basic argument of the present chapter. For three reasons, the following discussion will focus on the Middle East. Firstly—containing almost 50 percent of proved global petroleum reserves, and Saudi Arabia as one of the world's key swing producers—the region remains a crucial area of global oil production. Secondly, although some key OPEC members are not located in the Middle East (most importantly, Venezuela), American representations of the dangers of foreign oil have historically almost always referred to the "Arabs" in the Persian Gulf.[12] As a consequence, American cultural constructions of dependency have primarily been shaped by the country's relations with the Middle East. Thirdly, the Arab populations of the Middle East have long featured in the American mind as one of America's principal cultural Others. In order to grasp what it has meant for the United States to depend on this region for the supply of a commodity as vital as petroleum, American—Middle Eastern oil trade should be seen primarily as a cultural encounter.

Constructed Geographies and Oil Encounters

In political and academic accounts about oil, the term "foreign oil" invariably means that oil that is produced outside the territories under American jurisdiction, while "domestic oil" denotes oil produced within them. Oil's foreignness is seen as a straightforward and entirely uncontroversial question of material geography and state borders. By contrast, at the heart of the argument that not all imported oil is equally foreign rests the proposition that geography is not entirely a simple fact of nature but rather a "social and historical discourse which is always intimately bound up with questions of politics and ideology".[13] Such constructions of geography are an inevitable element in the creation of American national identity because imagining the nation always involves making assumptions about that which lies outside its borders.

Edward Said made the notion of constructed geography a centerpiece of his argument in *Orientalism*. There is, he argued, a universal practice in human societies of designating a familiar space as *ours* and, opposed to it, an unfamiliar space as *theirs*. This space may be characterized as an "imaginative geography" because it depends on "a very unrigorous idea of what is 'out there,' beyond one's own territory. All kinds of suppositions, associations, and fictions appear to crowd the unfamiliar space outside one's own."[14] Said emphasizes that there is a strong hierarchical element in this relationship between Self and Other. In the case of orientalist discourse, a rational, developed, humane and generally superior Western world stands above an Orient with the Arab world at its center, which is imagined as aberrant, undeveloped and inferior.[15]

Melani McAlister's study of US foreign policy toward the Middle East rests on a similar foundation. She espouses Michael Shapiro's concept of "moral geographies" to argue that Americans after 1945 constructed a "cognitive map suitable for the new 'American Century'", a map on which meanings were attributed to certain regions and cultural as well as ethnic groups, and which then shaped Americans' ethical, political and cognitive understandings of the world. "Moral geographies", she concludes, "are deeply historical and highly contested products, forged at the nexus of state power, cultural productions, and sedimented presumption."[16]

These concepts enable an examination of the geographical origin of America's foreign oil that is very different from those accounts that describe oil's foreignness only in terms of state borders. International movements of petroleum inevitably take place within space that has previously been constructed in certain ways. In other words, oil trade between the United States and the Middle East can be seen as a cultural encounter in which American views of the Middle East and the meaning of oil trade are mutually constitutive.

Both Said and McAlister discuss the 1973 Yom Kippur War and the Arab oil embargo against the United States in those terms. They are primarily interested in examining how the energy crisis impacted American views of the Middle East. Said writes that the crisis gave rise to new images of the Arab world. Arabs ceased to be camel-riding nomads and instead became oil suppliers who were characterized by the menace they posed to the American way of life.[17] McAlister argues that the events shaped the American perception of the Middle East as a location where US national interests were "vital but by no means secure". Like Said, she also observes that the embargo was "the beginning of what would become a staple of American popular cultural images: the greedy oil sheiks, with their hands on America's collective throat."[18]

These accounts of the energy crisis are compelling, but they also tell only half of the story. The oil encounter helped reinforce and alter American views of the Middle East, but how did such cultural images of the Middle East shape what it meant for the United States to depend on oil from this region? An interesting starting point for answering this question is contained in Natasha Zaretsky's brief examination of the cultural effects of the 1970s energy crisis in the United States. Drawing a link between the oil embargo and a domestic crisis of the US welfare system, Zaretsky argues that in both cases national debates revolved around "racialized notions of dependency".[19] Her argument is that dependency is a subjective condition; its meaning cannot be separated from widely held beliefs about the Middle Eastern oil sheiks on which the United States now relied for oil supply: greedy and lazy people whose newfound wealth was the undeserved result of accidents of geology. Zaretsky notes this almost in passing, but her argument nevertheless points to an important insight: if dependency is a "racialized" concept, this implies that the meaning of foreign oil dependence varies with the cultural images Americans hold about the countries of origin of US oil imports. Dependency, in this view, can have various meanings that exist simultaneously, sometimes complementing, sometimes contradicting each other. These disparate readings of dependency are created out of pre-existing constructions of geography—the places of origin of foreign oil and the peoples who control it. The Middle East and its oil, as will be shown below, have occupied a special place in this cultural discourse of dependency.

Foreign Oil Dependence as Interdependence

Based on the belief that some sources of foreign oil are secure and others insecure, American understandings of dependency have historically oscillated between two extremes. At the one end, there has been the definition of dependency as laid out in the previous chapters. Dependency was a condition that made the country vulnerable to the actions of other states, a critical political problem and a threat to US national security. At the other end of the spectrum, however, reliance on imported oil has also been described as an economically efficient, politically benign relationship with other states, a "normal" economic activity constituting a form of international division of labor that entailed gains from trade for all parties.

These two readings of dependency roughly conform to the two main positions in a long-standing academic debate in the social sciences about the meaning of interdependence. This is no coincidence, of course. A country's dependence on foreign oil simply constitutes one element in a larger relationship of energy interdependence between an importing country and some other country that exports oil. Any definition of dependency contains theoretical assumptions about the meaning and implications of interdependence. At this point, it is important to clarify these conceptual foundations.

Interdependence as a Concept

As an analytical concept, interdependence has a long and convoluted genealogy.[20] Essentially, Dale Copeland argues, the long-standing dispute between liberal and realist thinkers about the appropriate interpretation of interdependence revolves around how each camp approaches the costs and benefits of such a relationship.[21] Liberals tend to argue that entering into a relationship of interdependence entails efficiency gains; hence, liberals emphasize the opportunity costs of maintaining autarky. Highly interdependent states, they argue, are likely to behave peacefully in their international affairs in order not to jeopardize the benefits they derive from these transactions. The evolution of this argument can be traced from Adam Smith through Richard Cobden and Sir Norman Angell to present-day liberal international relations scholars such as Richard Rosecrance.[22]

Realists, by contrast, charge their liberal colleagues with ignoring questions of power and security; they tend to emphasize the adjustment costs that states incur should a relationship of interdependence be severed abruptly. Fear of such disruptions, they claim, incentivizes states to adopt a mercantilist approach to securing that which they depend on. John Mearsheimer formulates this rationale succinctly: national governments will

> struggle to escape the vulnerability that interdependence creates [. . .] States that depend on others for critical economic supplies will fear cutoff or blackmail in time of crisis or war; they may try to extend political control to the source of supply, giving rise to conflict with the source or with its other customers.[23]

Thinkers since Jean-Jacques Rousseau have argued that interdependence, in effect, spawns not international peace, but conflict.[24]

It is striking to observe how closely connected this academic dispute about the adequate conceptualization of interdependence has been with the empirical reality of US dependence on foreign oil. In *Theory of International Politics*, Kenneth Waltz details neo-realist concerns about the distribution of power and vulnerability arising from mutual dependence, using the 1973 oil embargo as an example. European countries—heavily dependent on Middle Eastern oil—were forced to "plead their cause or panic", while the United States "did not have to appease Arab countries" as its direct dependence on the Middle East was low.[25] The liberal point of view is covered by Richard Rosecrance who uses the same example to examine the power of interdependence: despite US threats to use military force in response to the embargo, peaceful economic solutions were ultimately pursued in order to defuse the crisis. Co-operation between the consuming and producing nations became the guiding principle, which allowed consumers like the United States to avoid a costly policy of self-sufficiency. Everyone gained from the arrangement.[26] In addition to such "pure" liberal and realist considerations, Robert Keohane, Joseph Nye and others have used US foreign oil dependence to discuss versions of the concept of interdependence covering the middle ground between these two poles.[27]

Defining Interdependence in the 1970s

The claim that US foreign oil dependence posed a serious threat to the nation hinged on the adoption of a realist understanding of interdependence. This view, as the two previous chapters showed, emerged fully after the 1973 oil shock. This process of assigning meaning to dependency went hand in hand with a larger academic and political debate about the significance of new patterns of global interdependence. After a brief struggle, a decisively realist reading of interdependence came to dominate the political debate. As far as the question of energy was concerned, a liberal, more benign interpretation of interdependence was marginalized. What is more, even the basic insight that energy dependency between the United States and the Middle Eastern producers might be mutual was routinely ignored in favor of describing the United States as essentially powerless in the face of unrestrained producer might.[28]

To be clear, the argument here is not that American policy makers should have adopted a different, more accurate or reasonable view of interdependence in the 1970s. However, as the following pages will demonstrate, a realist reading of energy interdependence became dominant only as part of a historically contingent debate about the nature, meaning and policy implications of new forms of energy interdependence. Most importantly, such a view was not applied consistently but only to foreign oil from some countries—above all, OPEC and the Middle Eastern producers.

Interdependence between the developed oil-consuming world and Third World producers was a new feature of the international petroleum order in the early 1970s. Interdependence, as Kenneth Waltz points out, is always a relationship between equals.[29] It was not possible, however, to describe the interactions between oil producers and consumers as a relationship of equals before the producing countries asserted full national sovereignty over their oil resources in the so-called OPEC Revolution of the 1960s and 1970s. More on this revolution will be said below. Here, it is important to note that these developments overturned an older energy regime in which there was no interdependence between producers and consumers of energy because the producers did not, in fact, exert a sufficient degree of sovereignty over the production and pricing of their natural resources. In the 1970s, energy interdependence was thus a new phenomenon. It was frequently seen as part of a larger emerging challenge to established patterns of international transaction by the Third World, culminating in demands for a New International Economic Order (NIEO).[30] Algeria's President Houari Boumedienne, whose country was both a driver of the OPEC Revolution and the initiator of the United Nations meeting at which the NIEO declaration was adopted, embodied the connections between changes in the oil market and challenges to the international order. As Francisco Parra writes, OPEC became "a sort of vanguard for the Third World in a struggle to redress what many developing countries regarded as an economic system rigged against them by the industrialized world".[31]

In these novel circumstances, the meaning of interdependence had to be determined before appropriate policies could be formulated. Henry Kissinger acknowledged this at the Washington Energy Conference in February 1974. "As we look toward the end of this century", the Secretary of State proclaimed, "we know that the energy crisis indicates the birth pains of global interdependence." The fundamental question from Kissinger's point of view concerned the appropriate policy reaction to this new situation: "Will we consume ourselves in nationalistic rivalry which the realities of interdependence make suicidal? Or will we acknowledge our interdependence and shape cooperative solutions?"[32] Such questions were being widely asked in the foreign policy community, and became entangled with debates about America's global role in the face of national decline and Cold War détente.[33]

An exchange of arguments on the pages of the influential *Commentary* magazine illustrates this debate well. Originally a left-leaning publication founded by the American Jewish Committee, it moved sharply to the right in the 1970s to became an important voice in the emerging neo-conservative movement. In the magazine's January 1975 issue, international relations professor Robert Tucker had argued that the only appropriate reaction to the oil crisis was the use of military force against the Arab oil producers. The whole affair, Tucker wrote, "is providing us with a supreme object lesson in the politics of interdependence". Dismissing as fantasy the liberal notion of "pacific interdependence", Tucker claimed that only a policy based on US power would effectively contain the new "egalitarianism" which, if permitted to develop, was likely to lead first to chaos and then to an "international system far harsher than today's, or even yesterday's, system".[34] Among the many critical responses to Tucker's realist position was a letter by his Harvard colleague Stanley Hoffmann, printed in the magazine's April 1975 issue. Rejecting a policy based on the use of blunt military force, Hoffmann argued that the United States should "entangle the 'newly rich' countries in the economies of the old ones" to give them an incentive for peaceful, non-disruptive behavior. "Would not the concern for a world system both less brutal and less unfair than that of the past", asked Hoffmann, "require that we seek solutions that maximize interdependence [. . .]?"[35] To this, Tucker replied by arguing the classical realist point of view: maximizing interdependence was pointless and dangerous, as "interdependence is in itself one of the reasons for conflict".[36]

American debates about oil imports are so deeply saturated with realist assumptions about interdependence that it may seem astonishing to find a fairly rich liberal tradition in American thinking about foreign oil. But Hoffmann's letter demonstrates that a liberal point of view could be applied to the issue. In fact, the question of whether oil imports should be seen, as with the liberals, as efficiency gains or, with the realists, as a threat to national security goes back to the very beginnings of the hydrocarbon society. In the 1920s, progressives like geology professor Charles Leith argued that reliance on foreign oil was desirable for reasons of economic efficiency.[37] Such views continued to shape the American oil discourse, coming to the surface, for example, in the 1950s during the intense debate about the Mandatory Oil Import Program.[38] Like many other government documents, Nixon's 1970 task force on oil imports pointed out that import restrictions came at the cost of efficiency and the promotion of world peace through open trade.[39]

Even in the run-up to the 1973 embargo, when OPEC had already assumed control of oil prices and begun to nationalize Western oil assets, a liberal reading of the new interdependence was adopted by some high-ranking American officials. Under Secretary of State for Economic Affairs William Casey, for example, offered the following assessment:

> The principal foreign policy implication at our becoming a major importer of oil and gas is contained in the word, "interdependence." Our natural resources, whether fossil fuels or ores, however immense, are finite [. . .] Above all we must recognize that we live on an increasingly interdependent planet and must work in harmony and cooperation with all others regardless of political, economic, and cultural differences.[40]

Proponents of the liberal approach—a position that McAlister has termed the "global managerialist view"[41]—were concerned about what they saw as a "dogmatic", "antagonistic" and "alarmist" debate in Washington before and after the embargo.[42] Whatever the merits of their arguments, a liberal approach to energy interdependence was soon marginalized at the expense of a realist reading of the country's dependency as described in previous chapters. Even stalwarts of liberal economics such as the *Wall Street Journal* found it impossible to argue in favor of continued energy importation. "Economically, our best choice would be to use whatever energy the economy demands and to purchase it wherever it is available", suggested the paper in January 1975. Yet, as the embargo and the extent of US vulnerability had shown, oil imports were not primarily a matter of economic policy but of national security. Increasing oil imports from the Middle East was clearly intolerable. "In other words", the editorial concluded, "for national security reasons the economically optimum choice is ruled out."[43]

Two Discourses of Energy Interdependence

The dominant view emerging in the wake of the embargo claimed that it was impossible to think of energy interdependence between the United States and the Middle East as a simple trading partnership. It was impossible because the Arab producers had chosen to use oil as a political weapon. They had nationalized the oil assets of Western companies, unilaterally raised the price of oil, cut back on production and implemented an embargo against the West for political reasons. All of this, it was said, made it practically impossible to see US dependence on foreign oil as anything other than a problem of national security.

Interestingly enough, however, an alternative view of energy interdependence survived. In the shadow of the dominant realist approach, a more liberal reading of energy interdependence as an economically useful and essentially benign condition continued to be applied to imports from some countries. The fault line between those two readings of energy interdependence corresponds closely to the notion of secure versus insecure sources of oil presented at the outset of this chapter.

The Middle East is the primary example of a source of US oil imports described in terms of the realist view of interdependence. Given that oil from this region has long been considered least secure by US policy makers, this is not surprising. At the outset of this chapter, Michael Klare was quoted as saying that oil's dangerousness is determined by the instability, hostility or dangerousness of its place of origin. Thirty years before Klare, Vice-president Rockefeller had argued, along similar lines, that while the rise of import levels was problematic in itself, "even more alarming" was the fact that an increasing amount of this oil was coming "from unstable Mideast sources".[44] The discourse surrounding the problems of US foreign oil dependence refers almost entirely to OPEC oil and, more specifically, to oil from OPEC's Middle Eastern producers. So complete is this identification that energy independence has frequently been understood to mean not freedom from oil imports but "freedom from OPEC".[45]

The notion that the dangers of foreign oil are primarily associated with "undependable" Middle Eastern producers, in turn, has made possible the formulation of a policy approach in which the United States continues to rely on imported oil while at the same time minimizing the adverse political effects arising from interdependence. In the early 1980s, the Reagan Administration formulated such a vision in its National Energy Policy Plan. Abandoning Nixon's unrealistic goal of complete national self-sufficiency, Energy Secretary Donald Hodel aimed for "energy non-dependence" in which importation continued where this made sense economically but without the associated political vulnerability to cut-offs and interruptions. Rather than decreasing imports across the board, the strategy coupled an insurance policy, in the form of the Strategic Petroleum Reserve, with a diversification of energy sources according to how secure they were. "We are less dependent on OPEC and Arab OPEC, specifically", Hodel announced in December 1983, adding, "Our largest supplier today is Mexico. Canada, the United Kingdom, and Venezuela follow. Saudi Arabia is well down the list."[46] The message was clear: less Arab oil, less insecurity. A more liberal view of energy interdependence was enabled by the diversification of energy supply away from the Middle East and toward reliable, friendly producers elsewhere. In a very real sense, therefore, the assumption that the dangers of foreign oil are geographically layered has provided a condition of possibility for an alternative, liberal reading of energy interdependence.

This liberal reading of interdependence is an important feature of American political debates about energy in the present. Consider, for instance, a paper entitled "Energy independence or interdependence?" by Pietro Nivola of the Brookings Institution. Observing that US policy makers have long argued in favor of reducing the country's dangerous foreign oil dependence, the author makes explicit the selective logic of these proposals by asking, "Which 'foreign sources of energy'? Presumably, we are not debating inflows from friendly places like Canada and Mexico." Nivola then proceeds to put forward the standard liberal case for a policy of comparative advantage and mutual gains from increased interdependence. Rather than vilifying imports from all sources, the United States should formulate an economically efficient policy of relying on secure supplies by some producers like Mexico and Canada. An appropriate energy policy, in this view, is to "seek deeper integration" through a liberalization of North American energy markets in the framework of the North American Free Trade Agreement (NAFTA).[47]

The administration of George W. Bush employed a similar logic of liberal interdependence when speaking about energy from some sources—most notably Canada. The 2001 *National Energy Policy Report*, for instance, noted how free oil trade with Canada *enhanced* US national security and entailed mutual gains.[48] President Bush, too, pointed out the positive economic and national security implications of a liberal energy regime in North America. "Canada is going to be the largest exporter of crude oil to the United States", Bush said in 2001, and then added, "That's good for our national security; it's good for our economy."[49]

In sum, two rival readings of energy interdependence have co-existed in American debates about foreign oil. There is the dominant, realist interpretation, applied primarily to the Middle East, according to which importation is associated mainly with costs: vulnerability to cut-offs and interruptions, as well as threats to national security and the American way of life. In the shadow of this representation, however, energy interdependence with some producers—above all, Canada—has also been described in liberal terms: as a benign economic activity, involving important economic benefits and mutual gains from trade.

Liberals and realists alike would tend to account for this divergence in terms of a sober analysis of costs and benefits. As for dependence on Canadian oil, they might argue, the risk that sudden adjustment costs will arise from energy interdependence is very low. By contrast, the benefits of maintaining such a trading relationship are very high. Hence, the language of the market-place obtains. In the case of the Middle East, however, the opposite is true as the risk of supply interruption outweighs the benefits of interdependence; hence, the focus shifts to security considerations. In this manner, the two readings of interdependence are inextricably entwined with the description of the geographical origins of foreign oil as secure or insecure.

On the one hand there is Canada, which for the past four decades has been represented as a "producing friend" in US policy documents. "The ties between Canada and the United States are close", explained one report, "and the two countries have already recognized their energy interdependence in, for example, electrical interconnections".[50] Moreover, when Canada became the single most important source of foreign oil for the United States in 2001 this was seen as beneficial by the Bush Administration, which emphasized the two countries' shared security commitments, geographic proximity, NAFTA and integrated pipeline networks as well as "long-term reliable supply relationships".[51] In the context of the early twenty-first-century boom in oil and gas exploration, even the concept of energy independence, traditionally a fundamentally national policy, is sometimes extended to encompass Canada. In discussions of the positive security implications of "North American energy independence", Canadian oil supply is essentially described as being every bit as safe and domestic as oil produced on US soil.[52]

The Middle East, by contrast, has traditionally and routinely been described in opposing terms. Chronic Middle Eastern instability and political volatility were well-established tropes in American assessments of the region well before the OAPEC embargo. It was the use of the oil weapon in 1973, however, which firmly established the widely held assumption that Middle Eastern oil was essentially unreliable and insecure. The "ease with which the oil valve is turned on and off in the Persian Gulf", wrote the *Washington Post* in November 1973, just demonstrated the "urgency of eliminating our dependence upon so instable and insecure a source".[53] Today, Middle Eastern oil is still routinely described in the same terms: as unreliable, insecure and prone to disruption and political manipulation. If anything, the rupture of 9/11 has increased the purchase of these views. One American observer after another points to the nation's chronic trouble with the Middle East. Here is just one example:

> Over the past 30 years, we have suffered Middle East supply disruptions caused by the Yom Kippur War of 1973, the fall of the shah of Iran in 1979 and Iraq's invasion of Kuwait in 1990 [...] How many times do we have to be hit before we pay attention? No one can foresee what might lead to a huge supply shutdown or whether the present attack on Afghanistan might trigger disastrous changes. A collapse of the Saudi regime? A change in its policy? Massive sabotage of pipelines? Another Arab—Israeli war? Take your pick.[54]

Defining Interdependence

Liberal and realist approaches to interdependence concur that assessments of the costs and benefits of interdependence are made on the level of individual states. That simply means that although interdependence is a relationship that involves at least two parties, what really matters is how policy makers in each state define it. Each state deals with its own dependencies.[55] How does this process work? Almost the entire existing conceptual literature on interdependence seems to be based on the assumption that policy makers objectively, and more or less accurately, calculate the costs and benefits of interdependence simply by observing the world. However, there is nothing in the idea of interdependence that would require us to base the concept on such a positivist and empiricist foundation. Conceding to liberals and realists that categories such as cost, benefit, risk, reliability or vulnerability lead to different definitions of interdependence does not necessarily entail the assumption that such evaluations are objective, valid and free from cultural influences.

Reconsider the cases of Canada and the Middle East to get a better idea what this means. Many observers have pointed out that, for the past three decades, Middle Eastern oil producers have essentially followed their rational economic self-interest by reliably supplying petroleum to the United States, even when their other political interests differed sharply.[56] Discussing the repercussions of the 1973 embargo, Giacomo Luciani concludes:

> The perception that Gulf oil supplies are insecure and unreliable is still based essentially on just that one decision, and persists notwithstanding the fact that since that time Gulf producers have demonstrated more than once that they are able to deliver all the oil that is required even in the presence of conflict in the region.[57]

To emphasize Canada's extraordinary reliability as a friendly supplier, moreover, is to omit the fact that Canada, too, had its own version of resource nationalism in the form of the National Energy Program (NEP), formulated in the wake of the 1970s energy crises.[58] Canada's participation within NAFTA's liberal free-market framework for North American energy interdependence was neither a natural nor an inevitable process, but was achieved laboriously through the dismantling of a policy that was protectionist, nationalistic and, some say, marked by anti-American sentiment.[59] In 1975, in response to a Canadian announcement that all oil exports to the United States would be cut off within ten years, the *Los Angeles Times* reported a widespread US sentiment that "the Canadians are no different from the Arabs".[60]

To be sure, the argument here is not that the Middle Eastern producers have historically, in fact, been more reliable than Canada. The proposition is that the process of defining interdependence is not a simple, rational, objective, unproblematic calculus of costs and benefits. Joseph Nye argues that the sources of interdependence are partially social. Economic interdependence, he claims, "has a high degree of social, especially perceptual, origin", such that changes in perception entail changes in the nature of interdependence.[61] What Nye sketches out in passing is an approach to interdependence based on what in international relations theory would be seen as a conventional constructivist approach. One could further develop such a concept of interdependence based on the constructivist approach offered by

Alexander Wendt. In this view, the meaning of interdependence is continuously (re)negotiated through interaction between the interdependent parties. Interdependence is what states make of it, as it were.[62] The American encounter with the Middle East, in this view, would appear as a realist relationship of security concerns and vulnerability that emerges, among other things, out of repeated American experiences with Arab attempts to use oil as a weapon against consuming countries in the Western world—in 1956, 1967 and 1973. At the same time, both Canada and the United States have been socialized through repeated instances of smooth energy trade into a world of liberal interdependence in which security concerns are muted and institutions are created to cement this mutually beneficial relationship. As a consequence, the economic benefits of interdependence are allowed to dominate perceptions on both sides.

While even such a thin constructivist approach would open up an interesting new vista for understanding the empirical reality of interdependence, it would perhaps also be too simplistic. Most importantly, if the assumption was that interdependence is defined out of perceptions of reiterated interaction alone, the new approach would not easily be able to capture the rich cultural context in which such encounters take place. When the United States began oil trade with Canada and the Middle East, US policy makers already held well-established images of Self and Other, which included cultural descriptions of their trading counterparts. These contexts can be captured by paying attention to the constructed geographies in which oil encounters take place. In other words, interdependence should be seen as a social construction that emerges out of extant cultural raw materials provided by descriptions of national identity. Because identity is relational, these descriptions inevitably contain assumptions about the foreignness of the Other. These varying degrees of foreignness, in turn, help to constitute the meaning of interdependence, as something between a realist relationship of danger and insecurity and a liberal relationship of benefits and prosperity.

Applied to the question of American reliance on the Middle East, this means that dependency is located in a realist discourse of security concerns because of pre-established cultural constructions of the Arab world in an American moral geography. It is not because a sober, objective analysis of costs versus the benefits of interdependence dictates this view, or because repeated contact with Middle Eastern oil producers has taught the United States that it is most appropriate to define interdependence this way. The meaning of American dependence on Middle Eastern oil cannot be separated from well-established representations of the Arab world as being radically different from the United States. Both in political and cultural contexts, Americans have tended to describe the Arab populations of the Middle East primarily as foreign, inferior, dangerous, irrational, incompetent, backward, untrustworthy, treacherous or, in short, "undependable".

It is not difficult to show how the new interdependence between the United States and the Middle East was interpreted in terms of such long-held views during the energy crisis of the 1970s. By way of example, consider the following two documents. In the November 1976 issue of *Harper's*, Emmet Tyrrell, founder and editor-in-chief of the *American Spectator*, published an article about US foreign policy in the Middle East that, while primarily addressing the Arab—Israeli dispute, described what the author saw as the fundamental fallacy of all US policies toward the region. Tyrrell wrote:

> Arabs are not like Philadelphians, nor are they like Frenchmen, nor even Israelis. This is why our tidy notions about how nations should behave look so hollow and ridiculous when applied to the Middle East. Arabs are religious fanatics devoted to a non-Western warrior religion [. . .] They have no democratic tradition, no tradition of contract law, and a civilization that is more decadent than Andy Warhol's [. . .] For our goody-goody Machiavellians to spend much time ensnaring such a non-Western, premodern people into a life of reason and dignity is as hopeless as trying to get a laugh out of the eternally sour mug of King Khalid.[63]

While Tyrrell's piece certainly is an extreme example, it is extreme primarily with regard to the striking amount of racist slurs contained in the text. As far as the underlying message is concerned, the article merely offers a particularly explicit example of widely held American assumptions about the Arab world at the time. According to these images, "the Arabs" were an underdeveloped, irrational people unwilling to or incapable of participating in the modern world. As Tyrrell explains, it is precisely because Arabs are not like *us* that we cannot expect *them* to behave according to modern, enlightened standards of morality and rationality. If this is so, and if interdependence is always a relationship between equals, how could anything like a benign, liberal economic form of energy interdependence between the United States and the Arab world ever exist?

The second example makes this dilemma even clearer. In a 1976 Senate hearing, Edward Luttwak, then a senior faculty member at Johns Hopkins University, addressed the problem of American dependence on Middle Eastern oil. Having argued that the "prosperity of Americans and the economic survival of core allies" could not be permitted to depend on potentially unfriendly regimes that were neither stable nor progressive, he was pressed by a skeptical committee chairman, Senator McGovern, to elaborate on this notion. Luttwak replied:

> It is absolutely preposterous to suggest that in the long term the entire energy basis of the Western economy can be predicated upon developing a relationship with Saudi Arabia. It is preposterous because of the nature of Saudi Arabia as a country, the nature of its society, the nature of its politics and the fact that it exists in the Middle East.[64]

Luttwak's point was that dependence on Middle Eastern oil posed a serious problem because of the people who controlled that oil. It was the *nature* of the Arab producers, not merely their actions, which made dependence on their oil an impossible risk.

These examples underline the idea that dependency can indeed be seen as a racialized concept, a cultural construction based on pre-existing American images of Self and Other. The foreignness of the producer determines the foreignness of its oil, and in this manner opens or forecloses the possibility of reading energy interdependence as a normal economic activity. The remainder of this chapter examines this argument more closely with regard to the Middle East.

The Middle East in an American Moral Geography

The Middle East began to figure prominently in the American mind long before oil ensnared the two places in an ever-closer encounter. Since the eighteenth century, Americans have imagined the region as a biblical landscape, a site of religious struggles from the Crusades to the Arab—Israeli conflict, as "the cradle of civilization", as a place of great wealth, magic and mystery, tyranny, decadence and, more recently, as a site where various vital national interests are threatened by anti-American extremism and terrorism.[65] Studying these cultural images, some authors have emphasized how—in addition to the personal experiences of merchants, missionaries, tourists, soldiers, oil men and diplomats—millions of ordinary Americans have encountered the Middle East indirectly through the medium of religion and culture.[66] Other writers have enquired into the role of social scientists, experts and the news media in creating "knowledge" of the Middle East for American audiences.[67]

The cultural assumptions that have emerged in this manner are closely tied to the establishment and maintenance of particular identity positions. In the twentieth century, Edward Said argues, American representations of the Middle East were pre-shaped by an older Orientalist discourse, according to which a primitive and irrational Orient contrasted sharply with enlightened Western modernity.[68] In this manner, the foreignness of the East helped to establish Western identity.

Yet, as McAlister rightly cautioned, the Middle East has not just been part of the "East" as a result of a binary division of the globe into Orient and Occident. Instead, the region and some individual countries within it emerged as distinct entities in the American mind during the twentieth century. Perhaps more than anywhere else outside the West, the Middle East came to be represented as a site where US national interests were at the same time vital and under constant threat.[69] Even more importantly, American images of the region were founded on what Said called an "essentialization of Arabs as basically, irrecusably, and congenitally 'Other'".[70]

Three important facets of this essentialization are of particular relevance to the question of oil trade. The first concerns the long American history of representing Arabs as intrinsically emotional, irrational, incompetent, untrustworthy, violent or treacherous. From Mark Twain's nineteenth-century account of the Holy Land to contemporary academic studies, cartoons and cultural texts, representations of Arabs have frequently emphasized the Otherness of Middle Eastern populations relative to Americans.[71] Edward Said once claimed that "virtually anything can be written or said" about Arab culture "without challenge or demurral".[72] Judging by the representation of Arabs in cultural texts such as Hollywood movies, this situation remains essentially unchanged today. The typical Arab, writes Jack Shaheen, "is what he has always been—the cultural 'other'". Having studied more than 900 Hollywood representations of Arab characters throughout the twentieth century, Shaheen concludes that only 5 percent of "reel Arabs" are depicted as normal human beings. The rest are portrayed as "brute murderers, sleazy rapists, religious fanatics, oil-rich dimwits, and abusers of women".[73] Unsurprisingly, after the events of 9/11, Hollywood has continued this practice throughout the first decade of the twenty-first century.[74]

Secondly, and closely connected to these racial and religious stereotypes, American representations of the Middle East have often revolved around the claim that these societies are backward and pre-modern. Apart from the apparent gap in material standards of living, the region's backwardness is supposedly evident in its traditional modes of social organization and kinship, pre-enlightenment political culture, relative absence of democracy and secular state bureaucracies as well as its lack of growth-oriented modes of economic development.[75] As such, American descriptions of the Middle East constitute a good example of what Anne McClintock calls "anachronistic space"—that is, geography populated by backward, atavistic, irrational peoples who are "inherently out of place in the historical time of modernity".[76] Most importantly, according to an influential view among US policy makers after World War II, the root cause for the region's underdevelopment was inherent in the character of its population. Arabs are "non-inventive and slow to put theories into practice" and "skillful mainly at avoiding hard work", stated a 1949 Central Intelligence Agency (CIA) psychological profile of the Middle East.[77] Similarly, in 1969, a State Department assessment argued that certain modes of social organization and traditional values were behind what was seen as "Arab resistance to modernization". The report concluded: "What modernization requires of the Arabs, in effect, is their de-Arabization."[78] The depiction of the region as atavistic and pre-modern has been linked with a related discourse about the modernizing power of Middle Eastern oil and America's role in bringing this transformation about. The notion of American benevolence in the region will be discussed in some detail below.

Finally, in an American moral geography the Middle East has been linked to threats to American interests. Historically, the origins of some of those threats have been located outside the Middle East—for example, Soviet attempts to gain influence in the region during the Cold War. More importantly, however, the notion of threat has been entwined with representations of the Middle East itself, such that threats to American interests followed logically from who the Arabs were. After World War II, for example, American policy makers routinely attributed what they saw as a threat posed by virulent Arab nationalism to the inherent fanaticism, hostility and irrationality of the local population.[79] Behind it all was Arab "violence, emotion and ignorance", as President Eisenhower put it.[80] Another illustration of this point, briefly discussed below, is the more recent debate about the Islamic threat, anti-American extremism and the supposed clash of civilizations.

This brief discussion of American views of the Middle East obviously cannot do justice to what is really a much more complex, contested and fluid discourse, but it suffices to establish that the Arab populations in the Middle East have long been subject to a process whereby the American encounter of difference in the region was translated into cultural otherness. Like few other places on the global map, the Middle East has come to signify pre-modernity, backwardness, fanaticism, violence and, most recently, a fundamental challenge to US security. Importantly, the foreignness of the Middle East extended to its oil, to the notions of US dependency on and interdependence with the region and to US policy approaches toward securing the flow of the region's oil.

Becoming Foreign: The Middle East and Its Oil

The Middle East has been described by Americans as foreign territory for centuries. The cultural difference of the region also extended to oil produced in the Arab countries of the Persian Gulf and Iran. However, compared to the long history of representing the Middle East as one of America's primary cultural Others, the foreignness of Middle Eastern oil emerged relatively recently. It is a product of the OPEC Revolution of the 1960s and 1970s. Arab oil had frequently been represented as American oil prior to this revolutionary sea change in the international oil regime.

The Cultural Appropriation of Foreign Oil

In 1928, two American oil companies, Socony (later Mobil) and Jersey Standard (later Exxon), managed to secure a significant share in the Turkish Petroleum Company in what is today Iraq. It was the first American oil business in the Middle East. Beginning with these first foreign explorations, the reserves of the American oil companies became subject to a process of appropriation. The goal was physical and economic control of petroleum production. But this process also encompassed what might be termed the "cultural appropriation" of foreign oil. This appropriation rested on the claim, put forward by policy makers and the oil industry, that Middle Eastern oil was really American oil, *our* oil. Not only was this oil in "American hands", as it was often put, but also the United States had a legal and moral right to it. The cultural appropriation of Middle Eastern oil served to overwrite the simple geographical fact that these reserves were located outside the territory of United States. It brought Middle Eastern oil into the sphere of the domestic. With the increasingly firm entrenchment of American oil companies in the Middle East in the decade after World War II, this notion acquired greater significance. During the energy crisis of 1973–4, the sometimes explicit, sometimes tacit claim that Middle Eastern oil rightfully belonged to the United States contributed significantly to the shock caused by the embargo decision. It was *our* oil, after all, that was being withheld by the rebellious Arab producers.

There are several distinct elements in this discourse of appropriation. Of primary importance is the fact that the business of developing Middle Eastern oilfields was undertaken by American (and European) companies which, on the basis of concessions, directly owned the oil reserves, determined production quantities and set prices. The role of host governments, in turn, was very limited.[81] As Simon Bromley observes, "Despite the formal independence of the relevant states, the regional hegemony of the United States together with the virtually complete dominance of oil capital over the oil concessions meant that Western control over Middle East oil [. . .] was complete."[82] More generally, at the height of the postwar petroleum order during the 1950s, the seven major oil companies, five of which were American, directly controlled about 90 percent of oil reserves and production in the Third World.[83] US national oil interests and the private commercial interests of the US majors overlapped to such a degree that a close partnership developed in which American companies acted, in Painter's words, as "vehicles of the national interest in foreign oil".[84] Policy makers in Washington had long attached great importance to maintaining a modus operandi regarding foreign oil reserves, which, while private in nature, was nevertheless organized around the principle of nationality. As four high-ranking cabinet members of the Coolidge Administration advised the President in 1926, "That American companies should vigorously acquire and explore such [foreign] fields is of first importance, not only as a source of future supply, but supply *under control of our citizens*."[85] When, after World War II, the center of gravity in world oil further shifted toward the Middle East, and to Saudi Arabia in particular, US policy makers highlighted the importance of the fact that this oil was "in American hands" and needed to remain there.[86] There was a widespread, if largely unstated, sentiment that American oil companies served as extensions of the nation overseas—an assumption that finally came to the surface when it was suddenly challenged during the oil shock of the 1970s. "The major oil companies which are supposedly the agents of US power abroad", one senator complained in response to OPEC oil price increases, "are actively aiding and abetting this economic warfare directed against the US and our allies in Europe and Japan."[87] Even sober academics like Morris Adelman concluded that "the oil companies are now the agents of a foreign power".[88] In reality, of course, American companies in the 1970s neither acted as simple extensions of the nation abroad nor as the handmaidens of OPEC.[89] But the widespread assumption that the companies firmly belonged to the US sphere of the domestic, inside and familiar had long helped overwrite the foreignness of the Middle Eastern oil reserves under their control.

The second element shaping the discourse of appropriation was provided by the notion that US oil companies abroad benevolently supported the local governments in their quest for modernization.[90] Behind this claim lay what Emily Rosenberg calls "the ideology of liberal-developmentalism"—that is, the notion that material progress was a function of universally applicable laws and that the United States had a divine calling to help other nations replicate its own extraordinary success.[91] As Robert Vitalis shows in his fascinating account of American oil business in Saudi Arabia, the oil companies actively promoted this message for their own ends.[92] In 1950, to give just one example, C. L. Harding, member of the board of directors of the Socony-Vacuum Oil

Company, delivered a speech to American engineers in which he praised corporate America's role in developing the Middle East:

> In an area long famous for its tales of wealth, magic, and romance, we are witnessing today an even more fabulous development, the magic that is taking place in the Persian Gulf [. . .] the magic of vast wealth accruing to countries which a short time ago had a meager income indeed.

It was oil, Harding said, that was behind this miraculous transformation, but not oil alone: "The development of this oil is an activity in which American ingenuity and American capital are playing a large part and of which all Americans can justly be proud."[93] To a significant extent, in effect, the region "owed" the development of both its oil and its economy to the United States.

A third theme within the discourse of appropriation, and one closely related to the narrative of American benevolence in the Middle East, revolved around the argument that American ingenuity, technology, capital, enterprise and labor were required to make oil work as an energy source. The reason was simply that the populations of the Middle East lacked the money and skills to do this themselves. This argument appears most explicitly in the case of Saudi Arabia, by far the most important American oil endeavor in the Persian Gulf region. In the discourse of appropriation, the empty and inhospitable deserts of the kingdom became a new American frontier, one to be conquered by American oil men. As in the American West, this frontier marked the boundary between the domestic, modern, civilized sphere of Self and the foreign, atavistic, uncivilized sphere of Other. In this sense, the Arabian American Oil Company (ARAMCO) did not simply drill for oil in Saudi Arabia, it brought the Dhahran oilfields on the Persian Gulf coast within the domestic boundaries of an American imagined geography. As a Chevron publication described the company facilities in Dhahran, "part of America has been set down amid the rock and sand".[94] Inside this "little America", status, privilege and work was organized according to a strict hierarchy of race and segregation of Americans and Saudis, which blatantly contradicted the company's claim of liberal developmentalism but which echoed the prevalent description of the Middle East in an American moral geography.

Vitalis has called these institutions a "Jim Crow system"; Ian Rutledge, an "apartheid-like regime".[95]

Even the US State Department recognized "the inescapable fact" that, to a significant extent, Middle Eastern oil had been removed from the political geography of the region. In 1972, a secret department report noted that

> the companies exploiting the concessions in OPEC today are *alien to the producing areas*. They got their concessions some time ago and they set up their establishments, frequently very large ones, some of which have had many of the aspects of an *extra-territorial settlement*.[96]

In a very real sense, these various claims and representations helped constitute Middle Eastern oil as American oil.

The OPEC Revolution

The OPEC Revolution ended the appropriation of Middle Eastern oil through private Western companies. Rather than constituting a sudden upheaval, this revolution was a conjunction of several developments that, beginning in the 1950s, gradually dismantled the postwar petroleum order. This regime had been defined by a "private oligopoly with close ties to the governments of the major consuming countries".[97] The almost total control over oil exerted by a handful of major oil companies had already been challenged in the 1950s, when real oil prices declined as independent producers started to compete with the majors internationally. The power position of the majors was further undermined by their "obsolescing bargain" with host governments over the terms of oil business.[98] Fifty—fifty profit sharing, first established in Venezeula in 1948, became a model and quickly spread to the Middle East. By 1970, the division of profits was roughly 70 to 30 in favor of the oil-producing states.[99] The 1971 Teheran and Tripoli Agreements forced the companies to make more wide-ranging concessions to oil-producing states on the issue of oil prices. Moreover, the foundation of OPEC in 1960 was important, even though few thought so at the time. Having started, as Fiona Venn writes, as a "purely defensive organisation" created in response to unilateral cuts in the posted oil price by the international majors, OPEC matured during the 1960s and began a more assertive policy, infused with Arab nationalism, which aimed at direct control of oil resources.[100] National oil companies (NOCs) were created to wrest control over operations from the Western majors, who were forced to retreat. At the end of this process, Western oil assets had been nationalized in almost all OPEC countries, beginning in Algeria (1971), Iraq (1972) and Libya (1973).[101] Equity in OPEC oil exploration controlled by the Western oil companies fell from 94 percent in 1970 to 12 percent in 1981.[102] By the end of 1973, OPEC was in control of both the production and pricing of its oil. Access to and control over Middle Eastern oil, deemed so important to the national security of the United States, could no longer be said to be in American hands.

Explicitly, the OPEC Revolution was a reaction to the foreignness and disproportionate influence of the international oil companies (IOCs) in a sector directly linked to the sovereignty of the producing states.[103] One influential driver of these changes was the emergence of producing countries as significant actors in their own right. As Venn puts it, "The once almost totally undeveloped producing countries of the Third World were now developing nations: they had created a technocracy, albeit a small one; they had embarked upon ambitious development plans which required constant and increasing inputs of capital."[104] The emergence of the Middle Eastern producers as developing economies, as significant players on the political stage was described in the United States as an awakening of the region.[105] Oil had made this transformation possible. As one author put it, "Finally, after years of backwardness, Allah's oil has given the Arabs an opportunity to partake of the twentieth century."[106] The appearance of the host governments as actors equal to the consuming countries had profound consequences for the nature of US dependence on oil from the region. Before the OPEC Revolution, the United States had simply depended on oil *from* the Middle East. This oil had been owned, controlled, developed and shipped by Western companies. Now, the United States depended on *Middle Eastern oil*. This oil could no longer said to be in American hands. It was "Arab oil", oil as foreign as the people who controlled it; as foreign as the places from which it was imported. In this sense, the encounter between the United States and the Middle East over petroleum did not begin in the 1920s, when American companies first began to work in the region, but only with the conclusion of the OPEC Revolution in the 1970s.

By the same token, the entire notion of energy interdependence hinges on this emergence of the producers as sovereign. As described above, interdependence is a relationship between equals and, therefore, it became an empirical reality only in the 1970s. "The new interdependence", wrote the *Los Angeles Times* at the time, was a relationship as troubled as it was inescapable; it was also, the paper noted, inseparable from cultural images held by each party:

> Separated by 6,000 miles and by even greater distances in terms of customs, culture, religion, and social values, Americans and Arabs have found it hard to shake off generations of ignorance and prejudice and adjust to the new realities that draw them together [...] The situation is made even more difficult by the American perception of the Arab world, which has been warped by distance, by cultural stereotypes and, until recently, by a genuine lack of interest in the whole subject [...] Nothing that happened either in the United States or in the Middle East before 1973 prepared the Americans and the Arabs for the two-way culture shock that resulted from the sudden emergence of the new relationship.[107]

It is important to understand that the emergence of the new interdependence and the end of Western physical and economic appropriation of Middle Eastern oil are mutually constitutive developments. It is equally important to note, however, that the OPEC Revolution did not immediately end the *cultural* appropriation of Middle Eastern oil in the United States. The continued significance of the assumption that Arab oil was really American oil explains part of the outrage in the United States about the behavior of OPEC and the Arab producers' decision to implement an embargo against the United States in 1973.

How did the cultural appropriation of foreign oil shape reactions to the energy crisis? McAlister observes that two distinct representations of oil emerged after the 1973–4 embargo in the United States.[108] On the one hand, there was an argument within the foreign policy community that sought to remove oil from the sovereign control of OPEC producing states—for example, by constructing oil as a "common heritage of mankind". In this view, oil was a commodity that belonged not to any individual nation but to everyone. Petroleum should thus be shared for the benefit of both consumers and producers alike.[109] Similar proposals suggested that oil be "internationalized" under the auspices of the United Nations.[110] On the other hand, and in stark contrast to this position, there was a representation according to which Middle Eastern oil was "a commodity that rightly belonged in American hands". Without the capital, technology and know-how of the American companies, the argument went, Middle Eastern states would not have been able to use their oil resources to modernize. In short, "it was 'our' activity that gave 'their' resources value".[111] At the logical extreme of this position was the standpoint that the Arab oil producers never owned their oil in the first place and, therefore, did not have a moral right to withhold it from the United States. As libertarian philosopher and owner of the *Reason* magazine Tibor Machan wondered, "We need to ask, however, whether the OPEC countries are in fact doing something with what is indeed theirs. Do the OPEC countries rightfully possess the oil?" Conclusion: "It seems to me clear that they do not." The Arab producers had freely admitted US business interests into their countries and then, through "theft and extortion", confiscated what these American companies had created.[112]

Underlying both these positions was the continued cultural appropriation of Middle Eastern oil. This perspective, which was further supported by the long-standing description of energy consumption as an American birthright, lay behind a strong sentiment that the United States had an implied right to obtain Middle Eastern oil—a right that was mirrored by an obligation for the producers to sell their resources at reasonable prices. Conservatives like William Buckley argued that the United States had "the moral right to force a supplier to sell it the oil it needs" in case they refused to.[113] In the context of the embargo, the debate about US dependence on foreign oil was replete with unspoken assumptions about the country's entitlement to the oil resources which it had developed for so long. But for the first time, the OPEC Revolution had enabled the Middle Eastern producers to refuse to play by the American rulebook. At the historical juncture when the meaning of the new energy interdependence between the United States and the Middle East had to be constructed, the producers had chosen to become, in Said's words, "disruptors of order".[114]

Disruptors of Order

There was a widespread sentiment in the US foreign policy community in the 1970s that the Arab producers, and the developing world more generally, had decided to break with the prevalent liberal international system led by the United States. When Said describes the oil exporters of the Middle East as disruptors of order, he refers to this sentiment. Stephen Krasner argues that this challenge to global liberalism under the NIEO banner was the logical consequence of Third World attempts to achieve a redistribution of international power.[115] Oil played a central role in these considerations. The oil crisis, Robert Tucker claimed in 1975, had raised the prospect of a world order in which "an orderly distribution" of global economic resources was threatened because "the principal holders of power among the developed and capitalist states may no longer be the principal creators and generators of order".[116] Observing a tendency in the United States to portray the developing oil-producing states as inherently opposed to the more established Western powers, Said concluded, "If these new nations are not creators and generators of order, they can only be disruptors of it." Furthermore, "they disrupt because all they are and can be as a group is inversely equal and opposite to 'us'".[117]

But what does it mean to say that Middle Eastern oil producers disrupted the US-led liberal oil order? The end of World War II had seen what G. John Ikenberry calls "history's most sweeping reorganization of international order".[118] The new system was characterized by Cold War polarity between the United States and the Soviet Union, and, simultaneously, by American efforts to build an open, liberal order among the Western democracies. Doug Stokes and Sam Raphael argue that, as the most powerful country, the United States pursued a dual strategy in shaping this postwar order. American leadership remained, first of all, a "distinctly national project" aiming to ensure the country's unrivalled material preponderance. At the same time, however, the United States also acted according to a "transnational logic" by creating a positive-sum, open international system "within which other core powers and agents of globalized capital (transnational corporations) [could] participate".[119]

Securing the transnational flow of energy has been a key requirement in ensuring the viability of this system. As Stokes and Raphael describe it, "The US has long acted to retain dominance over, and uninterrupted access to, the world's major oil supplies, principally in the Persian Gulf." This was not done to secure oil supplies for the benefit of the United States alone, however. "Instead, such interventions have been intended to facilitate integration of oil-rich political economies into the wider global economy, for the benefit of all major players within it."[120] American oil interests in the Middle East, in this view, concern more than the specific problem of US foreign oil dependence. These interests also include the defense of a structural kind of power over friends and rivals alike, which arises from controlling the flow of oil. Creating a liberal oil order has been read by some as a means of achieving an American hegemonic grand strategy.[121]

In order for this strategy to be viable, Middle Eastern oil first had to be made accessible. The oil resources of Saudi Arabia played a central role in this regard, constituting, in the words of the US State Department, "a stupendous source of strategic power, and one of the greatest material prizes in world history".[122] Integrating this oil within a US-led international regime was not easy. As Robert Latham argues, the postwar liberal order had boundaries. The most obvious one ran between the Soviet Union and the Western democracies. Yet, the newly independent states at the periphery of the system were not naturally part of the liberal 'zone of order" either.[123] After a complicated process of policy formulation in Washington, the issue was finally resolved by the late 1940s with the help of a series of Great Oil Deals in the Middle East, the most important of which enabled ARAMCO to tap into the immense reserves of Saudi Arabian oil.[124] Markets for this oil were created in Western Europe, where postwar reconstruction under the Marshall Plan was premised on a shift from coal to oil.[125] This arrangement also helped preserve US domestic oil reserves, which had been strained by the Allied war effort. The whole system was underwritten by British and American power in the Middle East. A decade after the end of the war, Bromley concludes, the integration of Middle Eastern oil into a US-led liberal zone of order was complete.[126]

The central point is this: when OPEC producers asserted their sovereign control over oil in the 1970s, they essentially undid their integration into this system. From an American point of view, they had stepped outside the liberal zone of international order and disrupted its smooth operation to the point where the collapse of the entire structure seemed possible. This goes a long way to explain the apocalyptic quality of American debates about oil in the 1970s and the frequent comparisons with the upheavals of the 1930s. More importantly, the notion that the producers had stepped outside the US-led liberal order further explains how the possibility of a liberal reading of US—Middle Eastern energy interdependence was foreclosed. The proposition here is that a truly benign form of economic interdependence was possible only *within* the liberal zone of order, not across its boundaries. After the OPEC Revolution, in effect, oil trade with the Middle East was rendered problematic in much the same way as benign economic interdependence with the Soviet Union had long been impossible in the context of the Cold War.

From the American standpoint, the Arab producers appeared as disruptors of order for two reasons. Firstly, they fell outside a liberal order by virtue of who they were. The possibility of liberal order, as Latham argues, is linked to the notion of liberal modernity itself—"a mode of fashioning and sustaining aspects of social and political existence", including liberal governance of the polity, the market principle and open economic exchange, as well as individual and collective rights.[127] Ikenberry, too, emphasizes the role of shared democratic values in constructing a liberal postwar system of institutions between Western countries.[128] As discussed above, however, Americans tended to describe the societies of the Middle East as the exact opposite of liberal modernity: as anachronistic space, inhabited by an atavistic, pre-modern society incapable of democratic rule. These notions now extended to the governance of Middle Eastern oil. How could a benign relationship of energy interdependence be expected if one party did not share the fundamental principles of liberal economic exchange and social organization? How could the oil sheiks ever be trusted to behave rationally and peacefully? As Robert Tucker asked: "Why should men be 'reasonable,' according to Western lights, when they have come so far and so fast by being unreasonable?"[129]

Secondly, OPEC members were portrayed as having deliberately stepped outside international order through their actions: the nationalization of Western assets, unilateral increases of the price of oil and the decision to use oil as a tool of foreign policy. High-ranking policy makers, as well as the majority of experts and journalists, in the United States portrayed OPEC's goals and policies as an illegitimate, artificial and dangerous interference in the smooth operation of the free oil market. Such an oil market, free from private or state-sanctioned constraints, had never existed, of course. Claims that OPEC's actions were illegitimate and morally reprehensible nevertheless tied in neatly with a long-established American narrative according to which limitations of energy supplies were necessarily "artificial".[130] As a cartel, OPEC constituted a prime example for such a man-made limitation of abundance. During the 1970s, it became commonplace in US policy and media discourse to point out that OPEC had created an "artificial scarcity" of oil in order to drive up prices and to maximize profits at the expense of the West. The problem with OPEC, as Vice-president Rockefeller defined it in 1976, was that "we are in a situation where the OPEC countries have acted on a political basis, not on a free market basis, to raise their price of oil in the world market."[131] In the United States, such descriptions could tap into well-established views about the corrupting power of monopolies and cartels. They led to the near-universal portrayal of OPEC as an illegitimate, hostile actor that needed to be broken up.[132]

Moreover, in addition to interference with a free market, the decision to punish the United States with an oil embargo seemed to show that the Arab producers did not see oil primarily as an economic commodity but as a political weapon. A few American commentators pointed out that free markets implied that anyone, including Arab oil producers, could choose whether or not to sell their products and to whom, but the overwhelming majority of US observers described the use of the "oil weapon" as a despicable, fully illegitimate act of blackmail, as economic warfare directed against the United States.[133] Outrage about the embargo was amplified by the assumption, discussed above, that the Arab producers owed the development of their oil to the United States. They were, in effect, under a moral obligation to make their oil available to consumers in the West at market prices.

All these elements combined within a view of the new interdependence between the United States and the Middle East in which the producers were described as active, powerful, threatening, hostile, unreliable, unwilling to let the market decide prices and unwilling to fulfill their implied obligations as providers of oil. Under these circumstances, the emergence of a realist reading of interdependence was probably inevitable. Gerald Ford explained concisely in 1974 why US dependence on the Middle East was intolerable:

"Sovereign nations try to avoid dependence on other nations that exploit their own resources to the detriment of others. Sovereign nations cannot allow their policies to be dictated, or their fate decided, by artificial rigging and distortion of world commodity markets."[134]

Middle Eastern Oil Since the 1970s

There is little evidence suggesting that the basic parameters of the American encounter with the Middle East over oil have been redefined after the historical juncture of the 1970s energy crises. This is not to say that nothing changed. But these new developments did little to undermine the purchase of views that equate both the nature and the policies of Middle Eastern oil producers with foreignness, threat and danger. There are at least three distinct areas where such fears crystallize today: discourses about resource nationalism, the curse of natural resources and the clash of civilizations.

Resource Nationalism

The phenomenon of resource nationalism and the way in which it has been processed in American debates about foreign oil have helped to cement the assumption that Middle Eastern oil reserves remain outside a US-led liberal international order. Resource nationalism in the context of oil can be defined as a political attempt by producing states to limit the operations of private IOCs and to assert national control over natural resources.[135] The claim that resource nationalism forecloses energy trade on the basis of truly liberal economic principles differs in some ways from claims about the cartel nature of OPEC. Contrary to widespread expectations in the 1970s that the organization's market power would permanently shape a new international system, OPEC, as Jahangir Amuzegar explains, "never acquired the exclusive power to set oil prices or restrict any country's access to oil supplies".[136] Owing to the increasing availability of non-OPEC supply after the 1979 oil shock, and the growth of spot and futures markets, OPEC's market share and role as price setter declined. As a consequence, concern among US policy makers over OPEC was relatively muted during the 1980s and 1990s. For some time, there were signs that the firm nationalization of Middle Eastern oil reserves that occurred in the 1970s might be reversed, or at least diminished. The NOCs, which had been the vehicle of nationalization of Western oil assets, came under increasing criticism in their home countries. They also lacked the capital and technology to tap into geologically more challenging oil reserves. Moreover, the debt crisis of the 1980s afforded the International Monetary Fund (IMF) and the World Bank an opportunity to pressure the Persian Gulf debtors into a "Washington Consensus"-style opening of the oil sector to privatization and foreign direct investment by private Western companies. As a result, but with the notable exception of Saudi Arabia, the Middle Eastern oil-producing states, which had long shut out the international majors, now began to open up their upstream oil business.[137]

Ultimately, this process of liberalization failed. Partially, this was due to political opposition to it in the Middle East. In the high-price environment of the 2000s, increasing production was also not an urgent priority for the producers. "Access to acreage in the region that had begun to open in the 1990s appeared to be closing or at least becoming available on much tougher terms."[138] This ties in with the observation that resource nationalism has been a cyclical phenomenon historically, finding expression in the way in which the IOCs were involved in the development of the oil resources in question. In the early twenty-first century, we have seen a new turn of the wheel in this cycle.[139] In effect, a very significant share of the world's oil reserves has been isolated from the circuit of the global capital flows and market-based investment decisions that characterize a liberal economic order. In 2008, the private IOCs only controlled 15 percent of total world supplies while more than 80 percent were in the hands of governments and their NOCs.[140]

US policy makers have long described resource nationalism as a major obstacle to the American national interest in maintaining a liberal oil order.[141] The twenty-first-century version of Middle Eastern resource nationalism is no exception in this regard. The producers' refusal to allow Western capital to enter their oil industries is seen as a serious threat to Western energy security. At the beginning of the 2000s, the EIA predicted an annual 2.2 percent increase in world oil demand through 2020, arguing that two-thirds of this additional oil would have to come from OPEC reserves.[142] In order for this to happen, Rutledge calculates, the Persian Gulf producers would have to increase their aggregate production by 108 percent and their share of world exports from 35 to 47 percent. He concludes: "Needless to say, such staggering increases would require massive investments in additional oil-producing capacity in the Gulf states."[143] However, as some oil experts point out, it may not be in the economic self-interest of oil producers to increase production in a high-price environment due to the fear of the adverse effects of oil windfalls. NOCs were observed to reduce investments, either in order to slow down the rate of depletion or because their governments captured so much oil rent that the capital for production-capacity increases was missing.[144] Based on such observations, American articulations of danger revolve around the prospect of insufficient oil supply from the Middle East. Resource nationalism is portrayed as a critical obstacle to the development of new supply, and thus to the future viability of the global economy. As one group of experts cautioned, "If political factors were to block the development of new oil fields in the Middle East, the ramifications for world oil markets could be quite severe."[145] In 2007, similarly, former deputy commander of the US European Command, Charles Wald, drew a link between resource nationalism and threats to American oil supply:

> Energy exporting governments don't need to resort to full-fledged embargoes to hurt the US and other importers [. . .] Governments in oil-producing countries can also constrain future supply through investment decisions that lead to long-term stagnant or slowing growth in production and exports, or even decline. Often enough, future supply destruction is the unintended or accepted consequence of an insistence on government control of natural resources. Currently, an estimated 80–90% of global oil reserves are controlled by national oil companies (NOCs), which are highly susceptible to being constrained by political objectives, even if these undermine long-term supply growth.[146]

Reflecting such concerns, encouraging the opening up of the energy sector in the Persian Gulf became a central tenet of energy policy under George W. Bush—for instance, in the administration's 2001 National Energy Policy.[147] Some argue, moreover, that the 2003 invasion of Iraq should be seen as military effort to achieve what neither the Washington Consensus nor the ebb and flow of resource nationalism had achieved: the re-integration of Middle Eastern oil into a free market system where oil investments could be made free from domestic political constraints.[148]

The Resource Curse

The Middle East also remains outside the zone of liberal modernity, and thus cannot be part of a liberal interdependence relationship with the United States, through the various mechanisms subsumed under the label of the "resource curse"—that is, the claim that resource wealth is not so much a blessing for developing countries as it is a source of economic and political trouble.

Oil, it is said, has accounted for tremendous material change in the Middle East, such that "a pre-modern social structure, based upon kinship, has been catapulted into the modern age".[149] In the 1950s and 1960s, many development economists predicted that oil and other natural resources would allow newly independent resource-rich countries to catch up with the industrialized West. This, together with Seymour Lipset's famous claim that wealth is causally linked to the emergence of liberal democracy, could be read as a prediction that oil producers, simply by bringing their oil to the market, would soon enough resemble their Western consumers, acquiring core liberal values such as liberal governance of the polity, respect for individual and group rights, and free domestic-market relations.[150] Yet, more and more evidence suggested that this was not happening on the scale predicted earlier and, in the early 1990s, the notion of the resource curse entered the academic debate.[151] An enormous body of literature has since challenged the idea that oil wealth was an advantage for producing countries and, instead, emphasized various adverse impacts of oil abundance on economic, political and social development in these states. Oil, according to this literature, appears to hamper sustainable economic growth, it can prevent industrialization, hinder democratization, cause bad governance and wasteful allocation of resources, fuel military conflict and internal strife and cement patterns of clientelism and patronage through rent seeking and capture.[152]

The insights contained in the notion of the resource curse are relevant for US—Middle Eastern energy interdependence for a simple reason. In the international energy order, the Middle Eastern countries perform a vital function as exporters of large quantities of oil on which consumers in the West and in Asia rely. But, as the resource curse literature argues, it might be this very activity that prevents the further development of the region toward liberal modernity. In turn, as long as the producing countries do not develop into liberal democracies, they will also remain outside the liberal zone of order. In other words, there might be a paradoxical mechanism at work here: the reality of energy interdependence between Western consumers and Middle Eastern producers can be seen as an obstacle to the inclusion of the Middle East into a liberal zone of order. As long as the region remains outside, however, a truly benign relationship of interdependence cannot develop.

Clashing Civilizations

An American reading of interdependence with the Middle East continues to be shaped today by descriptions of the region as a cultural Other. The twin issues of political Islam and terrorism illustrate how the power of these cultural images has perhaps even increased since the 1970s. Consider political Islam first. Since the 1970s and 1980s, many Middle Eastern societies have experienced a resurgence of political Islam, or Islamic fundamentalism as it is often disparagingly called in the West. Born out of "disillusionment with borrowed ideologies and cultures", these movements reject Western models of secular development and modernization and seek to restore religion's central role in society.[153] In such Western descriptions, the project of political Islam frequently appears as a rejection of liberal modernity itself. In effect, as George Joffé points out, the emergence of political Islam has served to reconstruct the Middle East as America's uncivilized, pre-modern antagonist Other—an image that can then be mobilized in neo-conservative narratives of existential threats to the West.[154]

The issue of terrorism is closely related to this narrative. Since the early 1970s, Palestinian terrorism had been a central theme in representations of the Middle East in the United States. However, it was only after the Iranian hostage crisis of 1979–80, when US embassy personnel in Tehran were held hostage for 444 days, that terrorism began to be represented as a threat to US national security. As McAlister observes, "The Iran story became the paradigmatic signifier of America as a nation imperiled by terrorism."[155] The event gave rise to significant changes in how Americans described the Middle East: Islam began to replace oil, Arabs and the Holy Land as the primary attribute of the region. Radical or fundamentalist Islam, in turn, was closely linked with terrorism in an emerging body of thought which saw anti-Americanism and anti-modernism as inherent features of Islam, and thus inevitably pitted *us* against *them*.[156] Bernard Lewis' 1990 article "The roots of

Muslim rage" is perhaps the best-known example of this train of thought. Lewis claims that inherent in Islam there is a "hatred" that "goes beyond hostility to specific interests or actions or policies or even countries and becomes a rejection of Western civilization as such, not only what it does but what it is, and the principles and values that it practices and professes".[157] Within this clash-of-civilizations discourse, in effect, the Middle East has acquired a new quality. It is no longer merely a location where American interests are vital but insecure. It has become the source of radical opposition to values central to the American national project, a site from which the most serious threats to US security emanate. The terrible spectacle of the attacks on the Twin Towers in New York in 2001 seemed to drive that point home in the most dramatic fashion.[158]

The cultural antagonism that is the centerpiece of these descriptions of the Middle East spilled over into definitions of US foreign oil dependence. This became evident, for example, in the way in which high-ranking government officials described dependency in the context of the war on terror. As President George W. Bush himself affirmed, "It jeopardizes our national security to be dependent on sources of energy from countries that don't care for America, what we stand for, what we love."[159] In the statements of neo-conservative speakers, cultural and political antagonism appeared as an important factor that greatly complicated American efforts to secure foreign oil supplies. "The good Lord didn't see fit to put oil and gas only where there are democratically elected regimes friendly to the United States", as Dick Cheney put it in 1998.[160]

The rise of a more antagonistic portrayal of the Middle East has helped maintain and even reinforce the dominance of realist readings of interdependence with the region in several ways. As the previous chapter showed, Islamic terrorism is now in itself seen as a threat to the flow of oil from the Persian Gulf, rendering American dependence on Middle Eastern oil more precarious than before. More importantly, this antagonism is said to increasingly shape American relations with key producing states. US policy debates about oil have long contained an assumption that dependency requires the nation to maintain friendly relations with the oil producers.[161] Since World War II, Saudi Arabia has been an important ally in this respect. The role of the kingdom further increased when the Iranian Revolution spectacularly destroyed the foundations of US co-operation with Iran, America's other major postwar ally in the region. In addition, Saudi Arabia has long been the largest exporter of oil; it sits on top of 20 percent of the world's oil reserves and, for the time being, is the only major producer with spare production capacity significant enough to mount a meaningful intervention in the global oil market. The United States has worked hard to preserve its special relationship with Saudi Arabia.[162]

Yet, the events of 9/11 and the role of Saudi citizens in the attacks caused a strong backlash against the country in the United States. "So now we know: The Saudi Arabian regime is no friend of ours", wrote Irwin Stelzer two months after the attacks. As he explained,

> The rulers of this land that sits atop 25 percent of the world's proven reserves of oil have been playing a double game. Even as the Saudi regime has accepted American protection and nurtured its longstanding relationship with Washington, it has also been playing footsie with the organization that murdered thousands of Americans in the World Trade Center and the Pentagon.[163]

As Stelzer and many others pointed out, Saudi oil remained central to the American and global economy. For good or bad, the sentiment was, the United States was caught up in a "lethal embrace" with a state that was commonly described as a "corrupt totalitarian regime at sharp variance with America's most cherished values" or as "terrorism's enabler".[164]

Cultural and political antagonism alone does not necessarily preclude the maintenance of a smooth relationship of energy interdependence. After all, the United States and Saudi Arabia have never shared a common perspective on the world. As early as the 1970s, we find Saudi officials expressing this view quite bluntly: "The relationship is not out of love. In fact, we cannot stand each other. The relationship is out of common interest."[165] In other words, cultural and political values did not matter as long as there was a concurrence of national interests. As long as Saudi Arabia needed the revenue from sales of oil, mutual dependence was roughly symmetrical.

It is precisely this concurrence of interests that was said to be increasingly threatened by the rise of terrorism and political Islam in the 1990s and 2000s. Some Western policy makers and experts pointed out that the revival of religion in the political calculus of the region potentially had significant implications for the question of Western oil supply. In the Middle East, as Øystein Noreng puts it, Islam constitutes the "normative basis of economic activity such as the extraction of petroleum and the use of the ensuing revenues". A petroleum policy based on Islamic economic principles would potentially favor leaving more oil in the ground over maximizing production, and thus revenue.[166] Like resource nationalism, an explicitly Islamic petroleum policy can therefore be seen as a significant obstacle to the continued expansion of global oil supply.

Moreover, the forces of political Islam and internal unrest in Saudi Arabia have been described as a threat to the ruling class in the Saudi kingdom and, thus, to the security of American oil supply. Religious and economic discontent with the policies of the royal family—evident in incidents like the seizure of the Grand Mosque in Mecca in 1979—have presented a serious challenge to the viability of US—Saudi relations.[167] This challenge, in turn, has been seen as a serious threat to US energy security. Should the Saudi Government be toppled by radical Islamists, there would no longer be a common interest to sustain a form of energy interdependence that has worked well despite the absence of shared values. As Stelzer points out:

> The Saudis can be counted on so long as the regime in place is one that needs the money from oil sales. But it is not clear that the valves would remain open were bin Laden's crowd, or its equivalent, to take over. To those folks, cash would be less attractive than injuring the United States: They seem, for instance, to prefer caves to palaces.[168]

In this narrative, the US—Saudi relationship is described as precarious because of the constant internal threat to the Saudi Government posed by political Islam. The Islamic revival is first equated with extremism, here represented by Osama bin Laden, and then depicted as motivated solely by opposition to the West. Further, since all Islamists wholeheartedly reject Western modernity and material progress (caves, not palaces), they would not need the revenue from oil exports. In effect, the very foundation of energy interdependence is said to have disappeared. The extremists are free to use oil as a weapon in their fanatical struggle against the West. Such statements contain more than a faint echo of the notion of American helplessness in the face of unrestrained producer power that characterized debates about foreign oil dependence in the 1970s. In 2006, former CIA Director James Woolsey made the same point forcefully during a congressional hearing: "It is often said that whoever governs the oil-rich nations of the Greater Middle East will need to sell their oil. This is not true, however, if the rulers choose to try to live, for most purposes, in the seventh century."[169] Here, too, there was no basis for any form of benign energy interdependence because the producers were said to hold no economic interests in oil or even in modernity at all. Woolsey then followed up on this by presenting a long list of radical anti-American extremist groups, culminating with the conclusion that reliance on the Middle East would continue to be a major problem for the United States in the future. Given the persistent purchase of these descriptions of the Middle East, how can a liberal definition emphasizing the benefits of relying on Middle Eastern oil ever emerge?

Conclusion

This chapter retells the story of America's oil encounter with the Middle East in order to make one central point: the meaning of US dependence on foreign oil is shaped by American descriptions of the foreignness of this oil. Foreignness, in turn, should not be seen as a straightforward question of material geography or national borders. The boundaries that determine whether oil is inside or outside, domestic or foreign, safe or dangerous are established by American cultural descriptions of the outside world. They are thus linked to constructions of Self and Other.

A comparison between oil imports from Canada and the Middle East makes it perfectly clear that, in this sense, not all foreign oil is equally foreign. Michael Hunt writes that US foreign policy discourse has historically been based both on the affirmation of national greatness and a layered hierarchy of peoples populating the rest of the world.[170] These layered cultural images of the outside world have shaped definitions of the problem of dependency in the hydrocarbon society. Radically different views of energy interdependence have emerged from radically different descriptions of the Other in an American moral geography. There is no simple Self/Other binary drawn along national borders in which all who are outside are represented as equally different, foreign and dangerous.

Oil from the Middle East has been far more foreign than any other oil because the region itself has been placed outside various cultural boundaries: that of Western liberal modernity, that of a US-led international order or simply that of Western civilization as Samuel Huntington would define it. In effect, US dependence on Middle Eastern oil becomes an unbearable threat to national security. Canadian oil, by contrast, has been described as domestic oil for all practical purposes, despite the obvious fact that it is foreign in the sense that it is outside US sovereign control. But Canada is not an Other to the same extent as the countries of the Middle East are. Canada is very firmly located within the boundaries just described, and thus not very foreign at all. In effect, US dependence on Canadian oil becomes a simple und unproblematic form of international division of labor from which all parties gain.

The insight that constructions of oil's foreignness do not depend on national borders could perhaps even be used to argue that not all oil produced on American soil has been described as domestic. Some foreign oil comes from within the United States. In 1975, attempts by Native American tribes to achieve a greater measure of sovereignty over natural resources on their land led to the establishment of the Council of Energy Resource Tribes (CERT), occasionally self-entitled "the Indian OPEC".[171] Questions over Native American identity, economic development and self-determination became entangled with the issue of how best to integrate these energy resources into the American economy. This, in turn, led to the articulation of a new "Indian threat" or, from the Native American's point of view, a white backlash. Here, too, oil had to cross boundaries drawn within a layered discourse of Self and Other. What is more, the articulation of threat, as one Navajo tribal chairman complained, "wipes out any need to feel guilt or concern: it provides a justification for taking our resources".[172]

As the Navajo tribesman had already hinted, assigning various degrees of foreignness to imported oil has had consequences for the formulation of concrete energy policies. A realist dependence on insecure, dangerous Middle Eastern oil necessitated very different courses of action to the liberal, unproblematic, even beneficial reliance on Canada's natural resources. The radically different truth effects of these two definitions of interdependence have been clearly reflected in

American policies to diversify oil supply away from the Middle East and toward greater imports from Canada and the Western hemisphere. They have also been reflected in broad policy strategies that have allowed flows of Canadian oil to be dictated by the free market, whereas, after 1979, Middle Eastern oil was increasingly safeguarded by the direct presence of the US military in the Persian Gulf. Exaggerating to make the point, one could say that different readings of interdependence led the US Government to conclude a free trade agreement with Canada while sending the Rapid Deployment Force to the Persian Gulf.

CHAPTER 5

FIGHTING FOR OIL

War was in the air in the autumn of 1990. In August, Saddam Hussein's Iraq had occupied Kuwait, thus plunging the region into crisis. In response, US President George H. W. Bush announced that such blatant aggression would not be tolerated and quickly deployed the US military to the Persian Gulf. Bush publicly described the operation as a purely defensive measure to protect Saudi Arabia, but many Americans did not believe him. As they saw it, the US military had been sent to the Middle East to prepare for war.

American public resistance to this alleged build-up for war, as the President experienced first-hand, was vocal and unruly. On 16 October, Bush appeared at a Republican fundraising dinner in Des Moines, Iowa in the run-up to the congressional midterm elections. Before he had even finished his light-hearted introductory remarks, three men suddenly rose from their seats and started heckling him. "Bring the troops home from Saudi Arabia", they shouted repeatedly. Bush was forced to interrupt his speech. Then, the three men joined their voices in a thundering, rhythmic anti-war chant, which they repeated over and over again for several long minutes until eventually police hustled them out and Bush could continue speaking.[1] That was hardly the end of it, however. The same chant would be heard throughout the country during the entire run-up to the first Gulf War in early 1991. It continued to interrupt political speeches. During a Senate debate on authorizing the government to use force against Iraq, protestors shouted it so persistently that the public galleries had to be cleared by the police.[2] Thousands of US citizens shouted it in the streets at peace protests, rallies, marches and vigils all across the country. It was scribbled on placards, posters and walls and generally became a rallying cry for those opposing the Middle East policy of the Bush Administration. It was always the same phrase, just four short words: "no blood for oil".

In American energy discourse, oil and violence have long been inextricably linked. Since the beginning of the hydrocarbon society in the early twentieth century, Americans have believed that the production, control and consumption of petroleum is tied up with the waging of war, the use of brute physical force and the violent death of innocents. As Amitav Ghosh once wrote:

> to a great many Americans, oil smells bad. It reeks of unavoidable overseas entanglements, a worrisome foreign dependency, economic uncertainty, risky and expensive military enterprises; of thousands of dead civilians and children and all the troublesome questions that lie buried in their graves.[3]

Saying "no blood for oil" is among the most common ways of articulating this link between oil and violence in the United States today. Yet, it is also a relatively recent addition to a much richer discourse of blood and oil, which can be traced back to the very beginnings of the hydrocarbon society. For instance, oil was tied to the spilling of blood in the places where it was produced. First in the American West and later abroad, oil companies and their workers clashed over questions of workers' rights, pay and representation, a struggle that frequently spilled over into strikes, sabotage and sometimes violence.[4] "Blood and oil don't mix" read the placards of striking workers and seamen invading New York's Times Square in 1939.[5] But oil has also been linked to war between states early on. Beginning with World War I, petroleum fuelled new military technologies and thus entered into the strategic calculus of military planners around the world. "Gasoline", wrote French Prime Minister Georges Clemenceau to Woodrow Wilson in 1917, "will be as vital as blood in the coming battles".[6] Oil became quite literally the lifeblood of the modern military machine. "Blood and oil title of next world tragedy", wrote the *Chicago Tribune* in 1923, predicting that another great war was coming. "It is going to be a war for oil and made with oil. For, once more, oil control means world control".[7] The claim that oil is very closely linked with political and military power has appeared crucial and self-evident to American statesmen and thinkers for a long time.[8] It has also featured in academic texts studying links between oil and internal warfare in producing countries or the possibility of resource wars between states.[9]

This final chapter examines the link between oil and violence as reflected in the no blood for oil slogan. In other words, it examines the claim that the United States Government might wage war against another country over oil; that half a world away blood might flow and innocents die because Americans require more oil. The chapter is interested in what American politicians, military men, security experts, oil men, journalists, academics, philosophers and ordinary citizens have had to say about the possibility, feasibility, necessity and morality of unleashing the awesome American military machine to ensure that the world's oil might continue to flow to those who want it. More specifically, the following pages focus on particular moments of crisis in the Middle East that triggered debates about war for oil back in the United States. These debates provided a framework for the expression of the US national interest in safeguarding American access to oil from this region.

There have been a number of such crises in the twentieth and early twenty-first century. The possibility that the nation may one day have to fight over oil was formulated in non-public US policy documents early on, well before the OPEC Revolution seemed to render Middle Eastern oil foreign and dangerous.[10] It was only after the 1973 oil embargo, however, that this question also became the subject of fierce and recurring debates in the US public realm. In 1975, America debated the military takeover of Saudi oilfields. Similar questions emerged once more in the wake of the Iranian Revolution. They influenced plans to create a Rapid Deployment Force, the formulation of the Carter Doctrine in 1980 and the eventual establishment of the US Central Command in 1983. They re-appeared during the so-called Tanker War from 1986 to 1988, when the US Navy intervened in the Iran—Iraq War to ensure the freedom of navigation in the Persian Gulf; in the run-up to the First Gulf War in 1990 and 1991; briefly after the attacks of 9/11 and again in the context of the 2003 invasion of Iraq. Most recently, the possibility of war for oil resurfaced over Iranian threats to close the Strait of Hormuz in 2012.

Three of these cases will be discussed here. The first is the debate about whether the US military should be employed to take control of Middle Eastern oilfields in the context of the energy crisis of the

1970s. Such considerations culminated in early 1975 in a truly public debate involving Americans from all kinds of social and professional backgrounds. The second and third case studies consider the two wars the United States and its allies fought with Saddam Hussein's Iraq in 1991 and 2003. In the run-up to both these interventions, oil figured prominently in public debates as a key cause for the waging of war. In both cases, however, these claims had an elusive quality, being rejected by some, affirmed by others and often likened to conspiracy theory. In the case of the 2003 war, most poignantly, the Bush Administration's official narrative categorically excluded oil as a reason for war, a claim that is still fiercely debated today. As in the 1970s, oil was inserted into larger discourses of morality and realpolitik as a basis for a military foreign policy.

The goal of the following discussion is not to assess the "true" role oil has played in shaping US foreign policy during these three crises. Whether or in what sense the invasion of Iraq in 2003, for instance, was really about oil is not relevant here. American debates about war for oil are worth studying because they represent moments of crisis in which the nation comes together to assess its relationship to oil, the material that constitutes the country as a hydrocarbon society. These debates have operated on the basis of pre-existing constructions of American identity, dependency, the culturally constructed foreignness of the Middle East and assumptions about the moral underpinnings of US foreign policy. Both those who argued in favor of war for oil and those who rejected such a policy were interested in the same questions, but found radically different answers to them. On a fundamental level, therefore, American debates about war for oil reveal how different courses of action could be depicted as legitimate, reasonable, necessary or unavoidable based on opposite readings of shared principles at the heart of the hydrocarbon society. To enquire into this discourse of blood and oil is to examine different, sometimes contradictory descriptions of American national identity.

Fighting for Oil after the 1973 Embargo

Suggestions that the United States should wage war against the oil producers of the Middle East did not appear in direct response to the beginning of the embargo in October 1973. Talk about military force had already emerged a few months earlier in the context of a Libyan threat to nationalize American oil assets and a Saudi Arabian warning that production levels might be reduced. In May, Senator William Fulbright advised the Arab producers not to push it too far. As the Arkansas Democrat warned, "The meat of the gazelle may be succulent indeed, but the wise gazelle does not boast of it to lions". Fulbright warned that the United States may "come to the conclusion that military action is required to secure the oil resources of the Middle East, to secure our exposed 'jugular'".[11]

The Nixon Administration quickly denied any such intention in public, but internally Kissinger and the new Defense Secretary Schlesinger were discussing contingency plans, among them an Iranian proposal for a joint intervention to safeguard Persian Gulf oil by pre-emptively invading key Arab producing states.[12] During a news conference in September, President Nixon himself then publicly issued a thinly veiled threat against OPEC. He ominously reminded the attending journalists what had happened to the Mossaddegh Government in 1953 after Iran had nationalized Western oil assets.[13] After the outbreak of the Yom Kippur War in October and the onset of the embargo shortly afterwards, the Nixon Administration increased the pressure. On 21 November, Henry Kissinger warned that the United States might eventually take "countermeasures" against the Arab oil producers if the embargo persisted.[14] Within the administration a preliminary schedule for war was even drafted, according to which intervention would begin in November when several American destroyers would be in the region for a pre-scheduled military exercise.[15] In January 1974, Schlesinger went on television to say that the producers "should not tempt fate by pushing the concept of national sovereignty too far".[16] In the Middle East, meanwhile, these threats were taken very seriously. Sheik Yamani, Saudi Arabia's oil minister, and others warned that production would be cut and oilfields blown up if the American military intervened.[17] Kuwait began to build minefields around its oil installations.[18]

Even after the end of the embargo in March 1974, the US Government maintained a scenario of military intervention in the face of what was seen as an ongoing threat to petroleum supply posed by OPEC. In August, Nixon resigned over the Watergate Scandal and was replaced by Gerald Ford. In September, the new President refused to rule out force in a major speech on the energy situation. "Throughout history, nations have gone to war over natural advantages", Ford said and affirmed that a sovereign nation like the United States could not allow its foreign policy to be dictated by the use of oil as an economic weapon.[19] Although the President argued that co-operation was required, his speech was widely taken as a threat. This impression seemed to be confirmed by visible military activities. An aircraft carrier, the *USS Constellation*, and several destroyers moved into the Persian Gulf on the first such visit since 1948. Officially, it was described as a "familiarization tour".[20] US military exercises were also held in the Mediterranean, with military personnel stating publicly that the United States was ready to invade Libya.[21]

January 1975 marked the apogee of official threats to use force against the Arab producers. So far, Kissinger had dodged the question of whether another embargo would constitute a *casus belli*. In an interview with *Business Week* published in January 1975, however, the Secretary of State suddenly stated what had thus far only been hinted at by officials. Kissinger declared that "military action on oil prices" was a "very dangerous course". On the other hand, he added, "I am not saying that there's no circumstance where we would not use force. But it is one thing to use it in the case of a dispute over price; it's another where there is some actual strangulation of the industrialized world". He immediately qualified this remark by adding, "I want to make clear, however, that the use of force would be considered only in the gravest emergency".[22]

Kissinger's "strangulation remark" caused a frenzy of public reactions, ranging from applause to strong criticism. It also forced the Ford Administration to take a stand on the issue. Schlesinger used a press conference to affirm that although the administration did not anticipate such an event, force would be used in the "gravest emergency", by which he too meant "strangulation".[23] The President upheld Kissinger's position and also made an effort to clarify what was at stake. "Strangulation, if you translate it into the terms of a human being, means that you are just about on your back".[24] The following day, Ford affirmed again that the United States had the "right to protect itself against death".[25]

Doubts have been raised about the seriousness of such threats. Much later, Schlesinger claimed that he had been "ready to seize Abu Dhabi" and that talk of war "wasn't just bravado".[26] Others argue that the US Government never intended to use force and that such saber-rattling was just psychological warfare to increase the pressure for a diplomatic solution.[27] In any case, the government upheld the war scenario until the debate gradually petered out in the spring of 1975. No shots were fired.

Favoring War

The question of whether unleashing the most powerful military machine in the world was an appropriate response to the intransigent behavior of the oil producers was posed not only within the US Government. From angry motorists hit by high gasoline prices to legislators on Capitol Hill, the issue was debated frequently and with great emotional force. As one senior official recalled, "The Naval War College was filled with Marine colonels walking around saying, we're going to put those Goddamned rag heads back on their camels".[28] The *Washington Post* reported "whispered suggestions that the United States must use the Central Intelligence Agency to overthrow the offending monarchs of the Persian Gulf".[29] Some commentators wholeheartedly endorsed the logic of strangulation as formulated by Kissinger. In an interview with ABC, Hugh Scott, the Senate Republican leader, said: "I can't imagine this country not doing whatever it needed to do, either economic or military, to permit itself to survive, and we wouldn't be worth a damn as a nation if we didn't".[30] Editorials appeared in various national newspapers, partially endorsing the use of force. The *Washington Post* found Kissinger's threat "useful", the *New York Times* observed that Kissinger was merely "stating the obvious" and the *Wall Street Journal* asserted that the threat to use force was a necessary policy instrument.[31] The *Los Angeles Times* concisely summed up what many professional observers of US foreign policy seemed to think:

> Obviously, no American in his right mind wants to land marines in Libya or Kuwait. But if actions by the oil cartel caused the collapse of the international banking system, if millions of Americans or Europeans were thrown out of work because oil was not available to keep the trucks, trains and factories running, you're damned right there would be consideration of military action. People would demand it.[32]

In his account of the crisis, Andrew Cooper argues that the push for war was led by the fledgling neo-conservative movement:

> Pressure on the White House to move decisively on the military front came first and foremost from the outriders of the conservative movement, the neocons who advocated boots on the ground in the Middle East to secure America's oil lifeline and the outright seizure of Saudi oil fields.[33]

These conservatives based their demands for military intervention on a number of assumptions, the foundations of what was then still an emerging neo-conservative credo. This included, for instance, the belief in the centrality of military force as an instrument of foreign policy and the conviction that the Middle East constituted a primary theater for American national interests.[34] But neo-conservative views on the appropriate response to the embargo also rested on a reading of the crisis that was saturated with references to the cultural foundations of the hydrocarbon society. This included a depiction of oil as America's lifeblood, dependency as a profound threat to the nation's security and the Middle East as America's cultural Other. As the writings of three of the most ardent pro-war conservatives show, a policy of force was rendered legitimate, reasonable, even necessary on the back of these assumptions.

The first of these was William Buckley, a well-known right-wing author, commentator and TV show host—and by then already a conservative icon. In October 1973, five weeks before the embargo, Buckley published an editorial in the *Los Angeles Times* that he called "The case for the use of force to get Arab oil". One month later, he followed up with another editorial that again made this case. Buckley claimed that his argument about war rested on "Christian social thought". Similar to a beggar who had the right to steal bread to avoid starvation, the West had "the moral right to force a supplier to sell it the oil it needs". Ultimately, this meant "that the Western military must, at the margin, be prepared to [. . .] land an expeditionary force in Libya".[35] The point was not to steal Arab oil, however. The laws of morality required that the oil be bought once the West was in control of the production facilities. Buckley declared: "A military expedition, aimed not at taking over Arab territory but at forcing the Arabs to export their oil at the market price, would be justified, under extreme circumstances, by the laws of nations".[36]

The second author, already mentioned in previous chapters, was international relations professor Robert W. Tucker. Unlike Buckley's fairly theoretical analysis, Tucker explicitly advocated the use of force in two lengthy articles, the first of which had initially been turned down by *Foreign Affairs*. In early 1975, both were finally published in the neo-conservative *Commentary* magazine.[37] Tucker deplored the absence of meaningful American threats to use force in response to the oil crisis—a case of "overcompensation" for Vietnam in his view. The oil crisis, with its escalating energy prices, had raised the specter of a breakdown of international order comparable to that of the 1930s. Tucker suggested military intervention as the solution. Much of his article revolved around the feasibility of intervention, the likely Soviet and European responses, and American public opinion. Like Buckley, Tucker argued that it was crucial to act not out of self-interest but to safeguard a larger objective: international order and the viability of "America's global enterprise". Once the oil had come under American control, it had to be shared. "Any military intervention that failed to provide an equitable allocation of the oil on a cost-plus basis would be seen, and with no little justification, as a raw display of American imperialism".[38]

The third, and most interesting, pro-war text was an article published in the March 1975 issue of *Harper's*.[39] It was called "Seizing Arab oil". The article is an extraordinary document for two reasons. Firstly, it was published under the pen name Miles Ignotus, Latin for "unknown soldier". According to the editors of *Harper's*, Miles Ignotus was "the pseudonym of a Washington-based professor and defense consultant with intimate links to high-level policymakers".[40] Intense speculation ensued as to who the author was; suspects even included Secretary of State Henry Kissinger. As it later turned out, however, the article had been written by Edward Luttwak, then working as an adviser to the Department of Defense. He had written it together with several like-minded Pentagon employees including Andrew Marshall, the head of the department's own think tank, the Office of Net Assessment.[41] "Seizing Arab oil" is extraordinary, secondly, for the extremely blunt, jaw-dropping brutality with which the authors advocated war against Saudi Arabia. No consideration was given to the morality and legality of war other than to dismiss such thinking as misguided. The article caused outrage in the United States. The American ambassador to Saudi Arabia, James Akins, went on television to say that whoever had written it had to be "a madman, a criminal, or an agent of the Soviet Union".[42] Even the Saudi Arabian Government convened in special meeting to discuss it.

The article began with the observation that the Western "policy of appeasement" toward the Arab producers had failed, and that "OPEC must be broken". The only way to achieve this was to have the Marines take over Saudi Arabia's oilfields ("an occupation of ten years and probably much less would suffice"). The text then painstakingly detailed the logistics and tactics of such an operation. 14,000 Marines, one or two battalions amphibious-landed, would take over the oil wells on the eastern coast of Saudi Arabia, preceded by a surprise attack by the 82nd Airborne Division in C-5 and C-141 jet transports flown in via Israel and escorted by air-refueled Phantom fighter jets. The result: "OPEC members will be faced with US control of Saudi oil reserves, which, if worked to the full, could put all of them out of business for fifteen years. At this stage, reason is likely to prevail, and production is likely to be restored". The authors dismissed the idea that the Soviet Union would be drawn into the conflict. The logic of mutual annihilation implied that Americans could open fire on their own clients "with near-impunity". Iran's position, by contrast, needed to be taken into account. "Why not then discretely ask whether the Iranians might be willing to 'protect' Kuwait—and, incidentally, appropriate their oil". True, the world would condemn the United States for intervention. Some would mean it, but Europeans and Japanese would "condemn, cry, and partake of lower oil prices with a sigh of relief". So would the American public. Soviet influence in the Middle East would increase, but so what? "Let the Russians have the influence, and let us have the oil". This oil should then be sold well below market prices but above production cost. Thus, supply for the industrialized West would be ensured, while part of the profits could be channeled to the poor developing world as aid. A newly created international organization would supervise all of this.[43] The authors ended with an apocalyptic prediction of the consequences of inaction:

> In a sober assessment, mindful of all the political costs and all the strategic risks, it can be done. It must be done. For if we do not do it, Project Independence will in fact be Project Isolation, with a somewhat impoverished America surrounded by a world turned into a slum. Almost everywhere, this would be an authoritarian slum, the product of utter hopelessness among the poor and mass unemployment among the former rich, all of us being forced to finance the executive jets of the sheiks and the fighter bombers of the dictators.[44]

Rejecting War

Threats about military intervention by US officials and right-wing commentators triggered a public debate that amounted to the first truly national conversation about blood and oil in the United States. Calls for war were mostly rejected. Unlike pro-war articulations, which primarily emanated from the neo-conservative camp, rejections of the logic of force emanated from all the key institutions shaping American oil and foreign policy discourses: the Democratic and Republican policy communities, the national news media, the oil industry and, importantly, the American public. Polls indicated that an overwhelming majority—81 percent of American citizens—categorically rejected the use of military force over oil.[45]

Criticism stemmed from various vantage points. One central argument concerned the legacy of the Vietnam War. Cooper observes that in the wake of the Vietnam disaster the public had lost the will to "back up threats with force".[46] Indeed, polling data suggests that Americans were less disposed to ponder the use of military force than at any previous time during the Cold War.[47] Expressing a widely held sentiment, a fellow of the Brookings Institution wrote that calls for armed force were the "irresponsible fantasy of extremists who had not learnt the lessons of Vietnam".[48] Others charged the proponents of a policy of force with jingoism, and saw in their proposals a misguided attempt to come to terms with the Vietnam trauma.[49]

Many within and outside the administration, moreover, argued that invading and operating foreign oil fields was militarily or logistically impossible. A feasibility study undertaken for the House Committee on International Relations warned that military intervention would likely fail. "This country could easily defeat OPEC armed forces in any given locale, while seizing oil fields and facilities, but preserving installations intact would be a chancy proposition under ideal conditions".[50] Likewise, the Pentagon conducted several studies on military options, including direct amphibious assaults. After reading one of them, the Chief of Naval Operations, Admiral Holloway, noted on the document that "there is little we can effectively accomplish in [the Middle East]".[51] Similar doubts were raised by oil industry experts familiar with the area. Thomas Barger, ex-CEO of ARAMCO, published a long article explaining why a successful takeover of Saudi Arabian oil fields was unlikely.[52]

Yet others pointed out the paradox inherent in US plans to invade Saudi Arabia, a country whose entire military apparatus had been

Figure 5.1 The Pentagon's Arab oil war planning division, 1975. Demands for military strikes against the Middle Eastern oil producers were rejected on several grounds. This 1975 cartoon pointed out the inherent paradox in a possible US military invasion of Saudi Arabia, a country that had long received substantial military assistance from the Pentagon.[53]

supplied and trained by the United States. This apparent contradiction at the heart of US policies toward the oil crisis became a recurring subject of political cartoons. Figure 5.1 is a case in point.

In addition to these criticisms, foreign policy professionals warned about the serious repercussions that military action would invite. In a forceful cable to the State Department, Ambassador James Akins cautioned that war would destroy the conservative anti-communist bloc in the Middle East; make all OPEC nations, including America's staunch ally Iran, hostile to the United States; drive raw material producers into the Soviet camp; undermine the global standing of the United States; isolate Israel; jeopardize the safety of Americans living in the Middle East; and leave NATO in a shambles. Akins called the whole debate "criminally insane", and demanded an unequivocal condemnation of the invasion idea by the government.[54]

American Identity and the Discourse of Blood and Oil

American oil culture and constructions of national identity played a crucial role in shaping the debate about war for oil in 1975. Both those who favored and those who rejected military intervention in the Middle East drew their arguments from the same pre-existing cultural raw materials. Almost the entire range of narratives and themes in American oil discourse discussed in this book so far was mobilized in order to argue either for or against war. The two basic positions within the discourse of blood and oil were structured as opposites around the same big questions about the role of energy in American society, the principles guiding US foreign policy and the essential nature of the United States as a nation. The question of how to react to the provocation of the embargo, in effect, became a debate about national morality and identity in times of dependency. As the following examples will show, the energy crisis forced Americans to engage in a debate about what kind of society they wanted to live in.

Consider first the issue of constructed geography and American cultural images of the Middle Eastern oil producers. As we saw in the previous chapter, representations of the Middle East as America's cultural Other were central to American views of dependency and the claim that the United States was morally entitled to oil supplies from the Middle East. In 1975, Buckley, Tucker and Miles Ignotus (alias Edward Luttwak) all based their pro-war arguments on a series of assumptions that emerged directly from such cultural constructions of geography. As discussed above, Tucker and Luttwak described the Arab producers as disruptors of international order, motivated not by reason and benign economic considerations but by fanaticism, greed and hatred of the West. This assumption of the region's quintessential otherness helped to make the use of force a reasonable and legitimate course of action. Military action was also permissible, in their view, because the United States had a moral right to Middle Eastern oil. Buckley claimed that it was legitimate to force the producers to sell their oil, and Miles Ignotus suggested that Middle Eastern oil should be forcibly internationalized under the auspices of some new international organization.

In much the same way, those who rejected a policy of force also explicitly rejected the cultural assumptions about the Arab world on which the neo-conservative discourse of blood and oil was founded. James Akins' angry cable to Washington is a case in point. Why the fixation on the Middle East, the ambassador wondered, and on Saudi Arabia in particular? After all, the Saudis were one of America's long-standing allies. Canada had also announced its intention to end all oil exports to the United States within a few years, so "the same reasoning that would lead us to occupy Saudi Arabia could more logically take us north of our border". What was the point of having lulled Canada into a sense of security for the past century, Akins asked with withering sarcasm, if the United States never intended to make use of it? Akins altogether dismissed the notion that the United States somehow had a legal or moral right to take someone else's oil.[55]

Many others argued along similar lines. As Clifford Hansen, the Republican senator from Wyoming, said on the Senate floor:

> If Americans reach the point where they are willing to ignore the general ethics by which we have learned to live, they might just as well talk about taking Canada's oil. It is no more right to discuss taking over an Arab nation, than it is to discuss taking over any other nation. The Arabs are owners of Arab oil, just as the Canadians are owners of Canadian oil.[56]

Proposals for war were "the height of arrogance and folly", according to one reader of the *New York Times*: "After all, it is *their* oil, not ours. Or does free enterprise stop at the water's edge?"[57] A reader of the *Los Angeles Times* expressed hope that "Americans are still idealistic enough to be horrified at the idea that our country would ever go in and take something away from the rightful owners".[58] Through such statements, Americans challenged the notion that Middle Eastern oil producers were somehow different from other oil-producing states. Irrespective of all cultural and political divides, the Arabs owned their oil and could do with it as they pleased. There was no American entitlement to their oil, according to this position, no implicit American ownership of Middle Eastern oil resources and certainly no moral right to unleash the US military to ensure that their oil would continue to flow. This underlines the fact that the debate about war for oil was also a clash of radically different American representations of the nature of the world outside US borders.

At the same time, the debate also contained a struggle over different descriptions of what the United States itself was like. Practices of self-representation along the lines of the lifeblood or addiction metaphors structured the discourses of blood and oil. Proposals of a policy of force were replete with body-politic metaphors, depicting society's relationship to oil along the lines of the lifeblood metaphor—that is, as an inherently natural, benign and inevitable form of dependency. The Ford Administration's frequent use of the strangulation metaphor is a case in point. The body politic required oil as the body required oxygen: it could not be any other way. Strangulation, not by embargo but by intolerable prices, amounted to denying that which the nation absolutely depended on; it was attempted murder and thus a *casus belli*, a lethal threat that legitimized self-defense by all means necessary. Others made similar claims with the help of different metaphors. Senator Fulbright spoke of "our exposed jugular", Buckley of the threat of "asphyxiation".[59] Robert Tucker likened the body politic to a man "bleeding to death", thus exchanging oxygen for (life)blood.[60] All these metaphors contained within them narratives about American dependency, the origins of the present crisis and threats to the American way of life. They implied claims about what was at stake (survival), who was to blame for the nation's predicament (the Arabs) and how to escape the crisis (military force).

In pro-war statements, moreover, American consumers and their way of life were depicted as hapless victims of a crisis that had been imposed on the country from outside. Because the crisis was foreign-made, simple changes in US domestic energy policy could not solve the problem. Tucker, for example, openly rejected consuming less energy as an alternative solution to the oil crisis. Even if there was a government with "sufficient authority and will to take draconian measures", he wrote, such a policy would lead to a sharp decline in standards of living and would thus be inacceptable.[61] Edward Luttwak was even more explicit about this. In a 1976 congressional hearing, he criticized the "puritanical economic" interpretation of the energy crisis, according to which shortages of energy represented a natural decline in resources and in which oil consumption was equated with guilt and "sinful abundance". In truth, Luttwak argued, there was an abundance of oil in the Middle East. Shortages were "manmade", artificially imposed by the producers in order to threaten the United States.[62] The "problem of dependence", therefore, was not connected to American consumption of oil but to the political shortage created by unreasonable and hostile producers. Interestingly, Luttwak then made an argument very similar to the one on which this book is based. He argued that "mere ideas" and "opinions" about oil had helped shape policy responses to the oil crisis. In Luttwak's view, the use of force was not considered as a response to the embargo because of widely-held perceptions about the sins of abundance. "Those notions serve as an anesthetic because while OPEC wielded the knife, they supplied the rationales for inaction".[63] Luttwak, by contrast, argued that only political power was capable of offsetting the threat of artificial shortages of America's lifeblood. As he wrote in "Seizing Arab oil", "There remains only force".[64]

Rejections of war were based on a different interpretation of the same question. According to this reading, the American way of life appeared not as the victim but the source of the nation's energy woes. Dependency was not read as benign and innocent but along the lines of the addiction metaphor. Proponents of this view criticized the use of the strangulation metaphor as a dangerous exaggeration. Senator Hansen spoke for many when he said this:

> To contemplate going to war to sustain the special dependency we have upon another peoples' petroleum for our high standard of living violates the moral concepts most Americans have [...] Such talk would seem to reflect a thinking that confuses the issue of national survival with the issue of maintenance of the present extravagant, energy-consuming lifestyle that most Americans enjoy.[65]

Hansen's Republican colleague Jesse Helms affirmed this view. "We will never strengthen our national character by fighting a war to defend only technological convenience and prosperous comfort".[66] These statements were based on the notion that US consumption of oil was wasteful and excessive—and thus anything but natural, inevitable and benign. The principal remedy for the situation was to change US energy policy. Tucker's article, for instance, was criticized on these grounds by readers of the *Washington Post*: "It is a sad day when a distinguished professor of international relations comes to regard war as less draconian than reduced consumption".[67] A letter to the *New York Times* contained this question: "Are we, as a nation that already uses far more than its share of the world's energy, threatening to seize still greater amounts in order that we may continue to squander it?" And another reader declared bluntly: "If you think that I am going to let my son go to war so that some clown can buy gas for his Cadillac, you have another think coming".[68]

The crisis seemed to reveal a fundamental dilemma between dependency and the moral foundations of US foreign policy. How could America's mission to be a force for good in the world—an aspiration flowing directly from American exceptionalism—be reconciled with the need to secure the flow of oil from foreign countries? The consumption of oil was essential to maintaining the American way of life. Did the crisis mean, therefore, that the nation would have to choose between this way of life and a morality-based US foreign policy? Did debates about war for oil show that the country was about to throw morality overboard in order to sustain its addiction to oil? Some thought so. As Senator James McClure said on the Senate floor, proposals for war were making "a mockery of the principles on which this Nation is based."[69]

Senator Hatfield, a Republican from Oregon, voiced such concerns explicitly. During the twentieth century, Hatfield argued, the socio-economic structure of the United States had been transformed into a kind of "superindustrialism fundamentally dependent on the exploitation of the world's natural resources". The drive to realize the "19th century dream of boundless plenty and wealth" had created strong pressures to acquire those resources wherever they were found and by whatever means necessary. From this point of view, Kissinger's strangulation remarks appeared as a manifestation of a mercantilist logic that flowed directly from the socioeconomic foundation of American society. Hatfield asked:

> How soon will it take the world to understand, then that the boundless material needs of advanced industrialism will turn the American democracy into a nation helpless before the dynamism of its own, economic tyranny? Our options will dwindle, until the only alternative is conflict and aggression.[70]

Ordinary Americans, too, were concerned. In response to the Miles Ignotus article, for example, one reader wrote a letter to the editors of *Harper's*, claiming that:

> Miles Ignotus is a most transparent nom de plume for Dr. Strangelove, who clearly inhabits the Pentagon, and will continue to demand more of the American taxpayer's, and the world's, resources until he is permitted to destroy himself and all of the rest of us.[71]

Two important points emerge from the discussion of these debates. First of all, both sides in the argument explicitly mobilized pre-existing cultural resources to make their respective cases. To be sure, these pre-existing articulations did not simply or unproblematically cause the formulation or rejection of a policy of force. The point is that a particular course of action for US foreign policy—the waging of war—was constructed as sensible, legitimate and necessary or, for that matter, absurd, morally wrong and pointless through the evocation of various well-established representations of oil and identity in the hydrocarbon society.

The second point concerns the question of discursive authority. The pro-war position emerged from assumptions about dependency, oil consumption and the foreignness of the Middle East that were very well established in the American oil discourse in the 1970s. Even so, demands for war hardly acquired any traction in American society. What accounts for this seeming contradiction? One answer is that the power of a particular mode of representation is always intimately linked with the authority vested in the social apparatuses within which it is articulated. Discourse has to be carried forward, amplified and made visible through material practices. It has to be inserted into institutions of power. Yet, none of this happened in 1975. Buckley, Tucker, Miles Ignotus and a few other hawks ultimately remained isolated voices even within the conservative camp of the policy establishment. They could neither mobilize the financial, institutional and human resources of the oil industry or other important groups, nor was their message positively amplified within the mainstream American news media. What is more, despite Kissinger's strangulation remarks, the supreme discursive authority of top US state officials was not used to articulate the crisis as a just cause for military engagement. This last point is of particular importance. As Susan Brewer writes, "When Americans are called upon to fight, they want to know why Americans must kill and be killed. They expect their leaders to prove that war is right, necessary, and worth the sacrifice". Usually, this meant representing war in terms of morality. Throughout the twentieth century, Brewer argues, US governments have developed narratives of war that equated the pursuit of US national interests with America's democratic principles. Fighting was said to be necessary to make the world safer, more prosperous and amenable to the spread of freedom and democracy.[72] Such narratives, in effect, served to reconcile potential contradictions between a morality-based and an interest-based foreign policy.

In 1975, the discourse of blood and oil was not made part of this important practice. While US policy makers did not rule out the use of force, neither did they make any attempt to insert the national interest in oil into the kind of idealist narrative traditionally used to justify war. As a consequence, pro-war arguments remained almost exclusively located within a discourse of realpolitik that most Americans have long found despicable. To paraphrase Stokes and Raphael, the discourse of blood and oil got stuck within a "logic of necessity" that was fundamentally at odds with the "logic of humanity" that Americans expected in the formulation of the nation's foreign policy.[73] As Jesse Helms argued on the Senate floor, "The moral impact upon this Nation of a war waged strictly for economic self-interest would be more destructive of our national fiber and character than the austerity and economic confusion which is presented to us as the other alternative".[74]

The Gulf Wars of 1991 and 2003

In the space of little more than a decade, the United States fought two wars with Iraq. In January and February 1991, Operation Desert Storm drove the Iraqi army out of Kuwait. In March 2003, Operation Iraqi Freedom saw troops occupying Iraq to end the regime of Saddam Hussein. Although President George W. Bush declared "mission accomplished" in May 2003, the war in effect continued until the end of his presidency and beyond. At home, both wars were accompanied by vigorous and controversial debates about the pros and cons of military action. The administrations of George H. W. Bush and George W. Bush each offered rationales for committing the nation to fighting

Saddam Hussein, reasons that were weighed, debated, applauded or rejected by the American public. In each case, the principal opposition to the President's narrative came from more or less institutionalized anti-war movements.

As in 1975, the discourse of blood and oil formed a major element of these debates. Apart from this basic similarity, however, the two Gulf War debates were very different from the 1975 intervention debate. In the 1970s, oil was the unquestioned motive behind pro-war arguments but no blood was actually spilled. In the case of Iraq, violence abounded but the role of oil remained elusive and controversial. In 1990, the elder Bush initially named oil as a major reason for war only to drop this claim later. In the early 2000s, the administration of his son consistently and categorically denied that oil had anything to do with the war. The two Gulf Wars nevertheless provide another opening to examine the evolution of the discourse of blood and oil. The aim here, it bears repeating, is not to determine the "true" role of oil as a causal variable in these conflicts—an issue about which the ensuing discussion is entirely agnostic. The question is what those cases can tell us about how oil has been made to function as an interest worth killing for, how certain descriptions of national identity have enabled or foreclosed the possibility of committing US forces to fight for oil, and what all of this says about America's relationship to the material that has constituted the nation as a hydrocarbon society.

Fighting for Oil in the First Gulf War

On 2 August 1990, Saddam Hussein's troops invaded Kuwait. Amid international condemnation, George H. W. Bush declared that Iraqi aggression against Kuwait would not stand. On 8 August, the President announced Operation Desert Shield—the deployment of US ground forces to protect Saudi Arabia. Sanctions were imposed against Iraq. In early November, two days after the congressional midterm elections, the administration changed policy and began to talk about "an adequate offensive military option"—the beginning of what would later become Operation Desert Storm.[75] Bush then set an ultimatum for Iraqi withdrawal from Kuwait for 15 January 1991. After heated debate, Congress passed a joint resolution authorizing Bush to use military force only three days before this deadline was reached, and, after Saddam Hussein ignored it, war finally began. Following a month of air strikes, ground forces of a US-led multinational coalition drove the Iraqi army out of Kuwait in just four days. When it was all over by 27 February, about 100,000 people were dead.[76]

During the build-up to war, the Bush Administration put forward a number of arguments to garner public support and justify its policies. In doing so, the President drew on the time-honored practice of equating American national interests with the interests of the international system at large. As Brewer observes, "The message that what is good for America is good for the world drew on the belief in American exceptionalism. This treasured myth claimed that the United States, as the world's morally superior nation, had special responsibilities and privileges".[77] Bush's stated ambition to create a "new world order" out of the crisis mobilized this comforting notion that America's national interests and its democratic principles were essentially identical.

The official narrative employed to justify military action revolved around two binary oppositions: good versus evil and legal versus illegal. Saddam Hussein was depicted as the "new Hitler", the epitome of evil, a ruthless dictator whose troops had "raped" Kuwait without provocation. This amounted to a clear violation of international law against which the entire world, not only the United States, stood united. If permitted to profit from its annexation of a sovereign country and the theft of about 10 percent of the world's oil reserves, Iraq was certain to become a formidable threat to other states in the Middle East and indeed to the entire world. Hence, the President affirmed, "our cause could not be more noble".[78]

Throughout the autumn of 1990, there were public debates about the appropriate course of action, dissident voices opposing Bush's claims, protests in the streets, rejections of the government's policies and accusations that the President was misleading the public.[79] All in all, however, Bush's claim that the country stood united against Iraq was not without foundation. Pollsters registered "extraordinarily high levels of public support" for Operation Desert Shield. One collection of polls, for example, found that a mean of 67 percent of Americans supported Bush's approach. Once the fighting began, approval figures rose even higher, often exceeding 80 percent.[80] In the *New Republic*, "just war" theorist Michael Walzer gave his blessings to military action as well. It would indeed be a "clean war", Walzer wrote, "so obviously just that one wants to see it fought".[81]

Oil was central to the public debate about war. From protesters chanting "no blood for oil" in the streets to heated debates in Congress or the statements of President Bush, arguments on all sides were saturated with the discourse of blood and oil. "Once again", as one journalist noted, "the word oil has become the most spoken three-letter word in America".[82] Despite oil's ubiquity in the debate, however, its actual role in the crisis remained controversial and elusive. The question was not *if* military action in Iraq was about oil—few people denied that it was—but in what sense the war was about oil. Depending on the answers to this question, another issue arose: was that cause worthy of the blood of American soldiers? As in 1975, the different positions within the discourse of blood and oil were structured as opposites based on different representations of national identity in the hydrocarbon society.

It had been President Bush himself who raised the question of oil early on. In discussing oil as a motif for war, Bush could draw on a long-standing presidential declaratory history of defining the protection of Persian Gulf oil as a vital national interest, dating back to the Carter Doctrine of 1980 and beyond. In the administration's narrative, oil appeared not only as a specifically American national interest but as a larger global issue worth fighting for. In effect, Bush removed oil from a national discourse of realpolitik and defined its protection as a US moral responsibility emerging from American exceptionalism. As Bush claimed:

> The stakes are high. Iraq is already a rich and powerful country that possesses the world's second largest reserves of oil and over a million men under arms. It's the fourth largest military in the world. Our country now imports nearly half the oil it consumes and could face a major threat to its economic independence. Much of the world is even more dependent upon imported oil and is even more vulnerable to Iraqi threats.[83]

The President's top aides, Secretary of Defense Dick Cheney and Secretary of State James Baker, frequently reiterated this narrative. The United States, the message was, was not acting in its narrow self-interest. A dictator like Hussein could not be allowed to control the oil on which the industrialized world depended so fundamentally.[84] There were two elements to this argument: firstly, control of oil would allow Iraq to blackmail consumers in the West; secondly, oil revenues would allow Iraq to dominate the region and perhaps to acquire nuclear weapons. Hence, the United States acted not only on its own behalf but on behalf of the international community.

In addition to this argument, however, President Bush also spoke about oil as a genuinely national interest of the United States. Moreover, this interest was said to emerge directly from the American way of life. This theme surfaced most clearly in the early days of the crisis, and in particular during a speech that Bush delivered at the Pentagon on 15 August 1990. The President said:

> We are also talking about maintaining access to energy resources that are key, not just to the functioning of this country but to the entire world. Our jobs, our way of life, our own freedom, and the freedom of friendly countries around the world would all suffer if control of the world's great oil reserves fell into the hands of that one man, Saddam Hussein.[85]

Given the centrality of oil in US society, the President argued, the annexation of Kuwaiti oil reserves by Iraq constituted a full-frontal assault on the key tenets of life in the United States. As in the case of the 1970s, the threat posed by the crisis concerned more than just the mere dysfunction of the US economy. It put into question "our way of life", the very existence of the United States as a hydrocarbon society.

This amalgamation of the two logics of realpolitik and morality in the early official narrative did not go unnoticed. The *Wall Street Journal* observed "universal agreement" that the United States had sent troops to Saudi Arabia to protect the integrity of world oil supply, and urged Bush to explain to the nation "that this effort unites the materialistic goal of oil and the idealistic goal of world order".[86] The *New York Times* claimed that "sophisticated citizens" easily understood that the "harsh realities" of US dependence on Middle Eastern oil made military action in the Gulf inevitable, while the less sophisticated needed to be educated about the necessities of realpolitik.[87] The continued urgency of the country's foreign oil dependence, argued another paper, had put the country in a "must-fight position".[88]

Opposition to the official narrative focused not so much on the administration's attempt to insert oil into an idealist discourse of American global responsibility but on Bush's argument that American blood needed to be spilled to safeguard the national interest, that is to say to maintain the nation's way of life. These challenges emanated from various directions, including the Democratic policy community and the news media as well as organized labor, environmental and peace groups. Thomas Friedman of the *New York Times* argued that the real American interests behind intervention in Iraq were "not quite so lofty" as the President wanted the public to believe. "Laid bare", Friedman claimed, "American policy in the gulf comes down to this: troops have been sent to retain control of oil in the hands of a pro-American Saudi Arabia, so prices will remain low".[89]

Just like the official narrative, this opposing representation of blood and oil was replete with references to the cultural building blocks of American society. Unlike the President, however, who represented the identity of the country and its way of life as something worth protecting, critics spoke about the same issue as the very problem underlying the whole crisis. If dependency was driving the nation into a war, America's own wasteful ways were at fault. As one reader put it in a letter to *USA Today*, "We have met the enemy and he is us". This self-description was based on the claim that the nation was "held hostage to Middle East oil" by its own automobile culture. Another reader suggested that a future memorial for the American war dead be in the form of an internal combustion engine and an oil barrel.[90] Such criticism mocked official claims about morality and global responsibility, and sought to unveil US foreign policy as a disguised pursuit of narrow self-interest.

In a scathing letter to President Bush, published in the *New York Times*, the father of a US Marine accused the administration of having watered down fuel-efficiency standards and energy-efficiency measures. Bush's belief in the magic of the free market simply meant that Americans would continue to squander the world's energy resources:

> And now you have ordered my son to the Middle East. For what? Cheap gas? Is the American "way of life" that you say my son is risking his life for the continued "right" of Americans to consume 25 to 30 percent of the world's oil? The "free market" to which you are so fervently devoted has a very high price tag, at least for parents like me and young men and women like my son.[91]

Criticism of the country's energy foundation also emerged from institutions of political power. When Congress debated the joint resolution authorizing the use of force against Iraq in January 1991, Senator Edward Kennedy, for example, claimed that the principal driving force for war was the American practice of wasteful energy consumption. This, he affirmed, was not a worthy cause for war:

> Not a single American life should be sacrificed in a war for the price of oil. Not a single drop of American blood should be spilled because American automobiles burn too many drops of oil a mile. Not a single American solider should lose his life in the Persian Gulf because America has no energy policy worthy of the name to reduce our dependence on foreign oil.[92]

Not long after the beginning of the crisis, a critical discourse of blood and oil also began to crystallize within an emerging institutionalized peace movement carried by blue-collar workers and the middle class. Groups like Out Now, the Coalition to Stop US Intervention in the Middle East or the National Campaign for Peace in the Middle East organized rallies, marches and vigils, sometimes gathering tens of thousands in New York, Washington, San Francisco and other major cities.

These groups used their often limited resources to challenge the official narrative—for instance, through advertising. An Out Now advertisement in the *Nation*, to take just one example, asked: "Must we trade body bags for oil? Why not Give Peace a Chance?"[93]

The peace movement was extremely successful in compressing its criticism of the Government into a simple slogan that quickly became the principal linguistic symbol of the anti-war position: "no blood for oil". Similar to the Big Ideas providing a direction-setting form of shorthand in official US foreign policy debates, the slogan drew on well-established themes in American oil discourse, in this case the figurative equation of blood and oil, to justify a political position. More specifically, the no blood for oil slogan attempted to set concrete limits on the pursuit of the national interest in oil. War, it said, was not a legitimate policy instrument.

The omnipresence of the "no blood for oil" slogan across the full social spectrum of the anti-war movement was due not least to its inherent ambiguity. As many observers point out, "It's difficult to know exactly what is being alleged when one is confronted by the slogan".[94] For this reason, it has sometimes been dismissed as "bumper-sticker rubbish".[95]

Saying "no blood for oil" was compatible with a straightforward pacifist position that rejected force as an instrument of policy for any reason. More importantly, it also entailed a critique of the official narrative along the lines presented above. What was dismissed, above all, was the logic of realpolitik according to which war was necessary to maintain a vital national interest in safeguarding the American way of life. "How many deaths per gallon?" was another way of phrasing the same criticism.[96] Polling data suggests that this was a very widespread interpretation of the administration's motives for war. In a poll conducted in August 1990, 63 percent of Americans said that troops had been sent to Saudi Arabia in order to protect "the flow and price of Middle Eastern oil".[97]

There was, however, a third meaning compressed into the "no blood for oil" slogan. According to this view, war was being waged not for the *national interest* in oil but for the *private interests* of the international oil companies, an argument that had been almost completely absent from the debate in 1975. The blood of American soldiers, it was alleged, was spilled for mere company profits. Protestors often chanted things like: "Hell no, we won't go! We won't die for Texaco!" According to one poll conducted by the *Washington Post*, about one-third of Americans held a more negative attitude of the oil industry as a result of the war. The paper attributed this change to the belief that the industry was somehow involved in it.[98] This criticism tied in neatly with deeply held views about Big Oil, according to which the oil industry was out to conspire with, corrupt and take over the democratic institutions of the State. It also chimed with a classical Marxist idea that saw government capture by private interests as an inherent feature of capitalism. This version of the "no blood for oil" slogan was thus primarily located on the far left of the political spectrum. One Kristen Hamel, the Workers World candidate for the US senatorial election in November 1990, stated that it would be a "war that only benefits the Pentagon and Big Oil", explaining: "The war being waged by the US against Iraq is only in the interests of Gulf Oil, Exxon, Mobil and Texaco, all of which have a vast oil empire in Saudi Arabia".[99] A similar argument, although much more guarded, was made by Democratic Representative Marcia Kaptur in the House of Representatives. The congresswoman opposed what she termed a "war for profits related to the control of Mideast oil".[100]

Another variation of this argument was presented by Noam Chomsky, a well-known intellectual and outspoken political observer, in

Z Magazine. He argued that the war was about securing American control over oil prices through the continued domination of the Middle East. According to Chomsky, "Iraq violated a fundamental principle of world affairs—that the energy reserves of the Middle East have to be firmly in the hands of US energy corporations and trusted US clients like Saudi Arabia's elites". It was in this sense that corporate power was involved in the war: "Now we fight Exxon's wars and anyone else's, as long as they pay the proper fees, either because they want to or, if necessary, because we force them to. Have gun will travel. Destination: a warrior state domestically and internationally".[101] From Chomsky's point of view, it was not deliberate government capture by Big Oil but the logic of American empire that forced the United States to fight. Corporations appeared in this narrative as vehicles of American imperialism. Bush's ambition to create a "new world order" was nothing more than the application of an old principle to a new situation.[102]

All in all, attempts by the administration to combine discourses of morality and realpolitik with regard to oil backfired spectacularly—so much so that John Greene, historian of the Bush presidency, later questioned whether Baker and Bush should ever have talked about oil in public, "for that was the aspect the antiwar activists pounced on".[103] Indeed, of all the arguments provided by the administration, the logic of realpolitik and national interests in oil resonated least with the public. In one poll, only 39 percent said "a major shortage of oil that threatens an economic recession" would constitute a legitimate cause of military action.[104] Michael Walzer speculated that it was the materialistic nature of the oil argument that constituted it as a "dirty" motive for war, and thus immoral in the eyes of many Americans.[105] The *Wall Street Journal* implied as much, urging the President "not to be embarrassed" to point out America's materialistic interests in Middle Eastern oil.[106]

Despite the unmatched discursive authority of the President, the official narrative about blood and oil did not, in sum, gain significant traction with the public. Bush reacted to this problem by adjusting his representation of the crisis in the autumn of 1990. When, as described at the outset of this chapter, one of his speeches in October was interrupted by protestors chanting "no blood for oil", the President replied: "You know, some people never get the word: The fight isn't about oil; the fight is about naked aggression that will not stand".[107] Moreover, when in November a journalist suggested that Bush had begun to "shy away" from the topic of oil, the President said that the main reason for American engagement in the Persian Gulf was "Iraq's brutality, aggression, and violations of international law" and that demonstrators asserting that oil was the sole reason for war were simply wrong.[108]

Whether this change conformed to the administration's actual interests in going to war in the Persian Gulf or whether it was intended, as others believe, for "domestic consumption" does not really matter here.[109] What does matter is the observation that the very proposition that the United States should fight for oil seemed unacceptable to a majority of Americans. The whole affair showed that petroleum could not easily be made to function within an idealist logic of American exceptionalism and global responsibility. Bush's early attempts to justify war within a realist logic of the national interest also collapsed under the weight of criticism encapsulated in the "no blood for oil" slogan. Furthermore, no one spoke in favor of sacrificing American lives for the private interests of the oil industry. "None of us wants to trade blood for oil" said Republican Senator James Jeffords in Congress when it came to voting for or against war.[110] In the course of the crisis, particularly after the shift from a wholly defensive Operation Desert Shield to preparing offensive military force, oil became a subject that verged on the unmentionable. It was removed from the official narrative and replaced with themes that emphasized global unity and US selfless leadership against Iraqi aggression. When the dust of war settled in 1991, a majority of Americans expressed agreement with the idealist discourse offered by the administration and concurred that the war had been fought over moral principles. The dirty, materialist discourse of blood and oil, it seems, had lost purchase with the public.[111] But just over a decade later, when the nation again pondered war against the Iraq of Saddam Hussein, it would return with a vengeance.

The 2003 Invasion of Iraq

On 20 March 2003, the United States invaded Iraq in order to overthrow the government of Saddam Hussein. The policy of containing Iraq, pursued in the 1990s by the administrations of George Bush snr and Bill Clinton, had increasingly come under attack from neo-conservative critics, who pushed for a more assertive US foreign policy in general and regime change in Iraq in particular.[112] In the wake of the 9/11 terrorist attacks, these neo-conservative voices gained purchase in the White

House and the Pentagon.[113] Their interventionist stance was reflected, for example, in the doctrine of pre-emption, articulated in the administration's 2002 *National Security Strategy*, according to which the United States reserved the right to strike first against terrorists and rogue states wherever necessary.[114]

In order to win the public's support for military intervention and regime change in Iraq, the administration of George W. Bush launched a massive and concerted effort in August 2002. Again, the official narrative emphasized that military action was taken not out of narrow national interests but on behalf of both the Iraqi people and the international community more broadly. Susan Brewer offers this useful distillation of the narrative:

> Portrayed as a war to make Americans secure and Iraqis free, Operation Iraqi Freedom was intended to expand US power in a vital region. Although presented to the public as a response to the 9/11 attacks, the war continued a decades-long strategy of policing the Middle East. In promoting it, the Bush administration constructed an apocalyptic narrative to persuade Americans to fight a war of choice. In Bush's version, the Hitler-like Saddam Hussein threatened the United States and all free peoples with WMD [weapons of mass destruction]. To protect civilization, the War on Terror had to be fought against the Iraqi leader and his terrorist allies, the "Islamo-fascists". Regime change would result in a democratic Iraq and a peaceful Middle East.[115]

Oil played no role in this narrative. It had occasionally appeared as a factor constituting the importance of Iraq in neo-conservative statements about the country.[116] When war plans became the subject of public debate in 2002, however, oil was mentioned only as a means of financing the postwar reconstruction of Iraq. The Bush Administration, as Michael Klare observes, "took great pains to keep from mentioning oil as a casus belli".[117] But oil was not simply absent from the official narrative, regarded as playing no role. Rather, it always lingered in the shadow of the administration's narrative as an unspoken allegation, as a dirty motive for war that was unworthy of the United States. Whenever the issue arose, the Government categorically denied that oil played any role at all. When the Future of Iraq Project—a series of government-sponsored meetings involving officials, Iraqi exiles and private stakeholders—convened in September 2002, a State Department official (falsely) claimed that oil was not even on the agenda.[118] "The United States is not fighting for oil in Iraq", said another White House official, David Frum, arguing that the United States was too rich to covet someone else's wealth.[119] Secretary of Defense Donald Rumsfeld told listeners on a CBS radio program that the war had "nothing to do with oil, literally nothing to do with oil".[120] In 2006, President Bush, after a direct challenge by veteran White House correspondent Helen Thomas, stated that allegations that the war had been about oil were "simply not true".[121]

While the aim here is not to determine the truthfulness of these claims, it is not irrelevant that many at the time believed, and continue to do so, that the United States invaded Iraq for reasons that remained unstated, a suspicion that was greatly fuelled by a continuous stream of revelations that the administration's material evidence to support its narrative had been questionable, misleading or—as a leaked memo from London's Downing Street put it—"fixed around the policy".[122] In an environment of great uncertainty and confusion, critics used every scrap of available information to make sense of the events. In May 2003, Paul Wolfowitz, an early neo-conservative who had advocated regime change in Iraq for years, stated that the administration had "settled on" the weapons of mass destruction argument for bureaucratic reasons, further fuelling suspicions that there was a wide gap between the official reasons for war and the real, hidden motivations.[123] A survey of the American community of foreign policy scholars shows, for example, that there has been broad agreement that the official narrative offered by the administration did not constitute the major driving forces behind the decision to go to war.[124] On the one hand, this has led to academic enquiries into how the government misled the public. On the other hand, a multitude of alternative, sometimes conflicting accounts of the reasons for war have been offered. More than a decade after the events, the Bush Administration's reasons for invading Iraq remain contested and mysterious.[125]

In sharp contrast to the official narrative, alternative accounts of the causes of war are full of references to oil as a driver of US foreign policy. In the survey mentioned above, for example, oil figures prominently as the second most often mentioned reason for war.[126] Numerous historians, foreign policy scholars, energy experts and journalists offer accounts of the events that largely follow the logic of realpolitik that Bush claimed had nothing to do with the war. In these representations, the long-standing national interest in oil appears prominently. Michael Klare portrays the war as an instance of what he calls the "strategy of maximum extraction", that is, the aim of boosting the oil output of the major Persian Gulf producers to satisfy global oil demand.[127] Ian Rutledge offers a very similar argument, according to which war was intended to free Iraq's untapped oil reserves.[128] Doug Stokes and Sam Raphael argue that the Bush administration sought the "transnationalization of Iraq". By this they mean the re-integration of its oil reserves and political economy into the liberal, US-led international order.[129] Noam Chomsky and others argue that the ultimate objective was to secure American hegemony through control of the oil on which the Western world critically depends.[130] By way of contrast, others do not believe that oil was an important factor. Daniel Yergin affirms that expanding Iraqi oil output was crucial to the world economy but claims that "this was far from a well-shaped—or well-informed—strategic objective".[131] Halper and Clarke, too, argue that the neo-conservatives pushing for war thought little about the country's oil resources.[132]

Those who do believe in the importance of oil like to point to six decades of official declarations describing the oil of the Persian Gulf as a vital US national interest. It is hardly credible, they say, that this time oil should have played no role at all in the considerations of the government. As *Time* put it, "The Bush Administration is up against a prevailing world view that the burden of proof is on the US to show that it won't exploit Iraq's underground riches".[133] Drawing on this widely held suspicion, Noam Chomsky asserted:

> For any person of minimal rationality, it's obvious that the United States invaded Iraq because it's right in the midst of the richest oil-producing regions in the world. But we're expected to believe that the US would have invaded Iraq even if it was producing just dates and all the oil supplies in the world were in southern Africa.[134]

Criticism of the official war narrative did not emanate solely from academic and intellectual circles but also from the re-emerging anti-war movement. A number of major rallies across the country culminated in a day of protests across the world that gathered as many as 10 million people, perhaps the single biggest anti-war rally in history. A study of the opinions of Americans participating in these events found that "the one opinion shared by almost all participants is that the United States wants to invade Iraq to secure its oil supply".[135] The Retort Collective, a community of California-based intellectuals who describe themselves as opponents of capital and empire, observed: "If there was a single political thread tying the anti-war mobilizations together, it was undoubtedly the refrain of 'No Blood for Oil.'"[136] Against the official narrative of American morality and benevolence, these protestors formulated two critical representations of blood and oil that exhibit striking similarities to the debate about the first Gulf War. As in 1990, criticism focused on the logic of national interests as well as the logic of government capture through private interests. As in 1990, moreover, these critiques emerged from certain understandings of national identity in the hydrocarbon society. These were almost identical to the assumptions that had guided the anti-war movement a decade earlier. Instead of examining them again in detail, just consider this excerpt from a newspaper article describing some anti-war placards shown at a smaller anti-war rally in Oregon.

> There was recognition of the role we, as American consumers, have played in our grab for something the Middle East has in more abundance than us: *Go to War for SUVs? Hell No. Ride a Bike.*
>
> And the way that our government and its industry backers are regarding all of nature, whether here or abroad: *How Did Our Oil Get Under Your Sand? Two Wars Per Gallon.*
>
> One sign said *No Blood for Oil*; another, *No Blood for George.* Were they redundant? One sign figured so: its photograph of Bush was coupled with the words, "Die for Oil, Suckers!" [. . .]
>
> But then, another sign explained that *Bush is The Best President Enron Could Buy.*

> There was the consciousness that war is distraction from what is needed: *War Is Not An Energy Policy.*[137]

While these statements reflect all the major themes discussed above, there also was an important difference to the protests over the first war against Iraq. This difference concerned the emphasis within the critical discourse of blood and oil. In 1990, the dominant anti-war argument had challenged the official position that securing the flow of oil was in the national interest and thus legitimate. An alternative claim, according to which the country was going to war for the private oil interests of American corporations, did exist but did not dominate the discussion. In the run-up to the war in 2003, however, the anti-war movement focused very much on the argument that the Bush Administration was acting on behalf of those private interests. No blood for oil essentially came to mean no blood for Halliburton, ExxonMobil, ChevronTexaco or ConocoPhillips.[138]

This argument was frequently based on the claim that the government of George W. Bush was, as one observer put it, "the oiliest administration ever".[139] Government critics argued that, besides President Bush, key members of the administration had close professional ties to the oil industry—most notably, Vice-president Cheney and (then) National Security Advisor Condoleezza Rice. This was frequently cited as evidence of the corrupting influence of Big Oil on Washington's policies. Moreover, critics noted that Halliburton secured a profitable contract to rebuild and maintain Iraq's oilfields, that US troops seemed far more interested in securing the Baghdad oil ministry than other public infrastructure and that Texas oil men were sent to advise the new Iraqi Ministry of Petroleum. According to this reading, the war had ultimately been about privatizing Iraqi oil for the benefit of the major oil companies.[140] The Retort Collective, for example, put forward a near-prototypical argument of government capture by the oil industry according to which "the White House now acted as the executive committee of the American Petroleum Institute".[141] Similarly, an activist associated with the Chicago Coalition Against War and Racism told the press that "they [the government] are more interested in continuing to build the bottom line for their corporate cronyism oil business, which is who they really represent, than they are in taking care of business for the rest of us".[142]

The allegation, in other words, was that private interests were taking precedence over the common good, and that people had to die in order to maximize the profits of a few greedy and powerful corporations. According to a variation of this argument, private and national oil interests had combined to push for war. In November 2002, for instance, unionized postal workers in the Detroit area printed an anti-war editorial articulating both these links between blood and oil at the same time:

> This war is about crude self-interest, and crude oil. Numerous sources have reported that the people Bush wants to put in power in Iraq will turn that country's massive oil reserves over to American oil companies. That, and a desire for military and political domination of the region, are the real motives for Bush's war.[143]

Since the early 1990s, the arguments linking blood and oil had been pushed further into the territory of the unsayable. In 1990, the administration of George H. W. Bush had removed oil from its official war narrative in response to anti-war protests. Oil had begun to verge on the unmentionable in official statements. By 2003, it had moved much further into the territory of the proscribed. As Alan Greenspan wrote in his autobiography, "I am saddened that it is politically inconvenient to acknowledge what everyone knows: the Iraq war is largely about oil".[144] Of course, linking blood and oil in a discourse of realpolitik had already been politically inconvenient in 1990, as the Bush snr Administration discovered. In 2003, however, even the suggestion that such a link might exist became itself subject to criticism, giving rise to a new theme in the discourse of blood and oil. Today, speaking about oil as the driver of US foreign policy is often associated with left-wing misperceptions of American foreign policy, radical critiques of the government or conspiracy theories. For example, oil-related challenges to the official narrative were dismissed by British Prime Minister Tony Blair as a "conspiracy theory", and "one of the most absurd" at that.[145] The conspiracy label now even extends to serious academic analysis. Scholarly activists like Noam Chomsky, for example, have complained that there is now a "taboo" against mentioning the role oil has played in US foreign policy toward Iraq. Chomsky argues that people are expected to believe in the official explanations for the war and, "if you suggest otherwise, you are accused of indulging in a conspiracy theory".[146] In the same vein, many American foreign policy scholars, surveyed about their personal opinion about the invasion of Iraq, asked to remain anonymous for "fear of being branded a 'conspiracy theorist.'"[147] This goes to show how oil has increasingly acquired the status of a dirty, illegitimate factor in US foreign policy. It was never easy or unproblematic to include oil in idealist narratives about the direction of US foreign policy. As the two Iraq-war cases suggest, however, it is now virtually impossible.

Conclusion

On a number of occasions, events in the wider world forced the American hydrocarbon society to consider the possibility of using military force to achieve its oil-related national interests. In other cases, public debates revolved around the question of whether oil was, in fact, a reason for war. Studying these debates, as this final chapter has done, reveals a fundamental tension between competing logics of realpolitik and morality in the discourse of blood and oil. On the one hand, we look back at a decades-long history of official declarations stating that securing the flow of foreign oil constituted a vital national interest of the United States, one that would at the extreme be defended by the US military. On this declaratory level, such statements often went unopposed. There was no public outcry when Jimmy Carter announced in his State of the Union address in 1980 that the oil of the Persian Gulf was so important to the United States that any outside attempt to gain control of the region "will be repelled by any means necessary, including military force".[148] Moreover, when Ronald Reagan actually sent the US Navy to ensure the openness of the Strait of Hormuz during the Tanker War of the 1980s, there was very little domestic dissent despite a not insignificant number of US casualties.[149]

On the other hand, when it came to actual large-scale warfare, this logic of realpolitik has been opposed by a majority of Americans. Of course, some have seen the issue as a simple calculus of blood and oil and found nothing wrong with a military oil policy. "Don't you agree", as one citizen put it, "that several GIs killed each week is a small price to pay for the oil we need?"[150] But the majority has disagreed with such views. Owing to a long national tradition of justifying military action in terms of American exceptionalism, morality, democracy, benevolence and responsibility, Americans reject the logic of realpolitik that seems to be inherent in war for oil. With the incomplete exception of the 1991 Gulf War, no US administration has seriously attempted to insert the national interest in foreign oil into such an idealist narrative. As we have seen, this does not mean that such a position has not been put forward at all. In 1975, arguments about a moral right to use force over oil were made in the neo-conservative niche within the foreign policy community. However, these views did not gain traction because they were not amplified by anyone with sufficient discursive authority. In particular, they remained largely absent from the narrative offered by the US Government. As the three case studies discussed here suggest, moreover, it has become increasingly difficult to represent oil in those terms. More and more, oil as a driver of US foreign policy moved into the realm of taboo, underlining what Amitav Ghosh once wrote about oil in the United States: "The history of oil is a matter of embarrassment verging on the unspeakable, the pornographic".[151]

The three cases studied here are also interesting because they show how the different positions about war for oil emerged from the cultural foundations of the hydrocarbon society. In particular, both positions were structured as opposites based on radically different answers to the same questions: What was the nation's relationship to oil? What were the principal values guiding US foreign policy? Was preserving patterns of domestic energy consumption as enshrined in the American way of life a cause worthy of the blood of American soldiers? Did the United States have a moral right to oil from the Middle East? Examining debates about war for oil underlines the fact that descriptions of Self and Other have never been uniform or unproblematic in the hydrocarbon society. Those who rejected war for oil, for example, often based their arguments on a definition of the problem of dependency in terms of the addiction metaphor and the nation's general reliance on oil. Why wage war against someone else when the real problem was "us?" Those who argued in favor of war, by contrast, espoused the logic of lifeblood and foreign oil dependence, and emphasized that the producers had no right to withhold essential oil supply from the United States.

One particularly interesting question in this regard is how future debates about war for oil will be shaped by recent changes in American descriptions of dependency. As shown earlier, the hydrocarbon society has recently begun to re-articulate its relationship to oil. The lifeblood metaphor has increasingly been challenged by the emergence of an alternative depiction of dependency as an addiction. No longer an essential and invigorating substance, oil is now portrayed as poison—a drug that drains the life force of the nation. In this context, it is not hard to see how oil also becomes a matter of embarrassment when it comes to declaring it a national interest. How can it be worth the blood of American soldiers and innocent civilians to maintain an addiction? In other words, in a society that increasingly defines its relationship to oil in these terms, fighting for oil becomes altogether unacceptable—more drug-related crime than the innocent pursuit of the national interest.

CONCLUSION

OIL AND AMERICAN IDENTITY

What is the nature of US dependence on foreign oil as a political problem? What can we learn about US society by examining the way in which Americans have approached this issue over the past century? These were the questions with which this book opened. Its central answer, which runs through every one of its pages, is this: US dependence on foreign oil is a profoundly cultural phenomenon, a specifically American experience, an historically contingent and constantly evolving discourse that has always had significance for the country above and beyond the narrow fields of US foreign or energy policy.

This book has been interested primarily in one particular link between the problem of dependency and the broader social context in which this issue emerged: the question of American national identity. In what ways has dependency featured in the never-ending American debate about the essential nature of the nation? How has dependency helped to constitute certain readings of identity in the United States while precluding others? In turn, which roles have American images of Self and Other played in the political processes of defining the meaning of foreign oil dependence? The previous five chapters each examined particular facets of these questions. Taken together, they offer three important answers. The first is that the meaning of foreign oil dependence has been constructed and reconstructed on the basis of pre-existing cultural raw materials, including, centrally, descriptions of American identity. At the same time, where Americans have disagreed about the most appropriate definition of dependency, those struggles have often revolved around competing narratives of identity. The second answer is that foreign oil dependence has been described as an existential threat to the key tenets of American identity since the energy crises of the 1970s. The necessity to import dangerous foreign oil was said to undermine the future viability of the American way of life; the nature of the country as a free, independent and exceptional nation; and even the very existence of American society in its present form. The third answer is that, although dependency has been described as a threat to American identity, the discourse of dependency has also served to affirm, stabilize and perpetuate certain well-established assumptions about America as a nation. This performative element of dependency has been particularly evident in the American declaratory tradition of proclaiming energy independence as an important goal of national energy policy. Consider these three links between dependency and identity in turn.

First of all, to say that dependency has been constructed out of pre-existing cultural raw materials is to say that its meaning is not simply given or determined by the consideration of objective and quantifiable variables. In order to understand the nature of dependency, it is not enough to count the barrels. Rather, dependency should be understood in its larger cultural context. As David Painter observes in his study about US foreign oil policy in the mid-twentieth century, "Definitions of national security and national interest were not shaped in isolation from the nature of the society they were meant to defend."[1] Writing about dependency based on a close examination of the nature of American society has been one of the main ambitions of this book. In order to do this, Daniel Yergin's concept of the hydrocarbon society has provided an excellent starting point.[2] The first chapter of the current volume expanded the concept to argue that in the United States oil is a social construction whose relevance for American life exceeds economic issues or material lifestyles and, fundamentally, includes big cultural questions about the mechanisms of progress, the nature of modern society, the key tenets of collective identity and the search for the good life. In the American hydrocarbon society, oil has become, as Peter Hitchcock writes, "a cultural logic in its own right".[3]

The twentieth-century energy transition that reconstituted the United States as a hydrocarbon society thus involved a twofold transformation of the black, viscous substance that is crude petroleum. On the one hand, the petrochemical and other industries transformed oil into all kinds of fuels, and industrial and consumer goods that became increasingly important to everyday life in America. On the other hand, the alchemy of American imagination transformed oil into all kinds of symbols that gave meaning to this new American life. In 2002, US comedian Jon Stewart offered a satirical hyperbole of all that oil has become in this process: "Oil! Black gold. Texas tea. Benzene derivative. Emitter of carbon dioxide as by-product of combustion. Chemical compound. Sticks to seagulls. Foreign policy nightmare. Death of us all."[4] Oil has indeed been all of those things in the United States. As black gold, it has symbolized the American dream of opportunity, social mobility and personal fortune for each individual American. On the level of US society, it has meant something similar as America's lifeblood: an invigorating substance intimately connected to progress, prosperity and the good life. But oil also came to be equated with destruction, decay and catastrophe. It has been described as a drug to which America is hopelessly and slavishly addicted, a poisonous substance that drains the strength of the nation. In the context of a looming climate crisis and the specter of a world where natural resources are depleted, oil has been linked to catastrophes on a global scale—a claim captured in Stewart's exaggerated description of oil as the "death of us all".

Political definitions of US dependence on foreign oil emerge from cultural raw materials in the sense that they are shaped by such pre-existing assumptions, narratives and social relations, all of which form part of American culture. Not all of these extant factors are directly connected to oil. Some, for instance, relate primarily to the general process of defining political problems in the United States. In American political culture, the authority to explain the world and define a situation or event as a political problem is not evenly distributed. Some speakers, such as state officials—and, above all, the President of the United States—are vested with special meaning-making powers. In this view, dependency emerges from cultural raw materials because it is constituted within a particular set of social power relationships that is in itself an institutionalized social construction.

The cultural nature of dependency is best evident in connections between American descriptions of national identity and the meaning of dependency. It has been argued throughout this book that oil has been so central to life in the United States that it has seeped into the process of making and re-making American identity. And since such constructions of national identity always consist of assumptions about Self and Other, it is not surprising to see that different definitions of dependency have echoed variations, contradictions and changes in how Americans have imagined both themselves and the outside world. The book discussed two major areas in which different readings of identity have historically translated into different definitions of dependency. One is the recent debate about whether the problem of dependency refers primarily to American reliance on oil at large or rather to the country's dependence on imported oil. The other concerns the fault line between liberal and realist readings of US energy interdependence, and, therefore, the question of whether oil imports are a normal economic activity or a serious threat to national security. How are each of these questions connected to culture and identity in the hydrocarbon society—for example, through the cultural discourse of abundance? As discussed in Chapter 1, the experience of material abundance of energy sources has been an important part of American energy history, so much so that abundance has become a cultural discourse in its own right. In the early hydrocarbon society, the thin, cold fact that plentiful reserves of coal, oil and natural gas could be found on American soil was inserted into powerful narratives about energy consumption as a prerequisite for economic growth and progress, as well as individual opportunity and social mobility as encapsulated in the American Dream. Oil abundance was imagined not as an objective fact but as the result of American ingenuity, American political freedoms and the magic of the free-market model. This discourse constructed abundance as the normal state of affairs and oil resources as essentially limitless.

Prior to the 1970s energy crisis, definitions of dependency emerged directly from within this discourse of abundance. As Chapter 2 showed, dependency did not really exist as an urgent political problem before the 1970s. This had nothing to do with oil's material predominance over the American economy or lifestyle patterns. In 1960, oil supplied 44 percent of total US primary energy consumption, more than at any time before or since. The country's utter reliance on petroleum for the most basic daily activities was widely recognized but did not appear as something policy makers needed to worry about. Society's relationship to oil was described in terms of the lifeblood metaphor: oil was a healthy, benign, even invigorating element of the body politic. In this manner, the discourse of abundance helped to preclude the emergence of a reading of the country's reliance on oil supply as a political problem. In the context of abundance imagined as limitless, why think of oil reliance as a problem? This situation only changed when the American experience of the energy crisis in the early 1970s delivered a shock to the discursive foundations of the hydrocarbon society, including the discourse of abundance. Only when deeply held cultural assumptions about energy consumption, limitlessness and progress were ruptured, in other words, did some space open up for a redefinition of the country's oil reliance as a political problem. According to this new representation, the nation was guilty of squandering a precious and scarce natural resource, and, in the process, risked destroying the natural environment.

As Chapter 2 also showed, however, this new definition of dependency was quickly crowded out by a different reading of the crisis in which reckless foreign rulers were portrayed as the principal culprits and American consumers as their hapless victims. Dependency, in this view, was a political problem primarily in the form of the country's dependence on insecure oil imports. This interpretation became dominant because it was backed by the supreme discursive authority of the US presidents of the 1970s and their administrations. But it also fitted in neatly with the long-held American assumption that abundance was natural while shortages were artificial. OPEC, and the Arab producers of the Middle East in particular, were said to have deliberately interrupted the flow of an otherwise abundant commodity in order to punish the US Government for its foreign policy during the Yom Kippur War. Dependence on foreign oil, in effect, became the most important energy-related policy problem in the United States. In sum, as a cultural raw material, the discourse of abundance helped determine the meaning of dependency in two ways: first by foreclosing the possibility of a definition of dependence encompassing all forms of oil consumption before the 1970s, and then again by providing a background against which meaning could be given to the energy crisis.

The second area in which dependency and identity intersect is the debate about whether US dependence on foreign oil is a mutually beneficial economic relationship or a threat to national security. In this regard, as Chapter 4 discussed, American images of the outside world have provided a collection of cultural raw materials out of which disparate definitions of dependency were forged. The foreignness of foreign oil and, in effect, its dangerousness, emerge from ascriptions of cultural and political similarity or difference. In this view, US imports of Canadian oil are unproblematic because American policy makers have long seen Canada as a "producer friend": a stable, friendly, liberal country with many important similarities to the United States. The Middle East, on the other hand, has been described in radically different terms. The region has often been imagined as situated outside the zone of liberal modernity, in a pre-modern, atavistic space full of irrationality, backwardness and hostility. The most important effect of these radically different representations of the suppliers of America's imported oil has been to constitute some foreign oil as dangerous and some as safe. Middle Eastern oil became foreign and dangerous in a sense that could never be applied to Canadian oil. Emphasizing the otherness and unreliability of the Middle Eastern producers negated the possibility of reading the emerging energy interdependence between this region and the United States in liberal terms after 1973. The foreignness of Middle Eastern oil meant that dependence on this region was always and inevitably a problem of national security. Reliance on Western suppliers like Canada, by contrast, has been interpreted as a normal and benign form of mutual dependence involving gains from trade for each party.

A final observation regarding cultural raw materials: if the problem of dependency is created out of resources that are offered in the context of the hydrocarbon society, this means that dependency is a specifically American experience. Most industrialized Western economies have long been net importers of oil, and some of them rely on foreign supply for virtually all the oil they consume. Yet, in these countries the fact that oil has to be imported is not rendered meaningful in the same cultural contexts as in the United States. In many ways, the problem of dependency is thus a uniquely American issue. In countries like France, Germany or Italy, there seems to be no similar tradition and no comparable history of speaking about oil imports as an issue that is directly related to the identity, sovereignty, independence and power of the State or the viability of some European way of life. At the very least, such debates seem to be far less common, far less urgent and more confined to specialist policy circles in Europe. All of this is despite the fact that Western Europe has historically relied much more on supposedly dangerous Middle Eastern oil than the United States. The 1973 embargo, Yergin writes, had dramatic effects on the "psyches of the Western Europeans", suddenly putting into question their tremendous economic achievements after 1945.[5] Yet, the crisis does not appear to have caused a discursive shock in these countries comparable to what happened in the United States. Like the United States, the European importers joined the emerging consumers' oil regime under the auspices of the International Energy Agency (IEA) in 1974 and established strategic petroleum stockpiles. Afterwards, however, the main focus in European energy debates shifted to other questions such as Russia and the security of natural gas supply. This relative absence of urgency with regard to foreign oil dependence in Europe cannot be easily explained simply by reference to the American security umbrella under which Europe has enjoyed protection since the days of the early Cold War. Why should America's power be capable of providing energy security for Europe but not for itself? In order to understand the nature of dependency in these countries, we would have to examine the cultural contexts in which dependency has been described there.

The second link between dependency and American identity discussed in this book concerns a particular reading of threats to US national security. In the wake of the 1973 oil embargo against the United States, policy makers, journalists and security and energy professionals articulated a series of threats that they said resulted from American dependence on oil produced by undependable and hostile producers in the Middle East. At the heart of this discourse sat the claim that foreign oil was connected to grave dangers to the national security of the United States. Chapter 3 offered a close reading of the arguments put forward in this regard. It showed that there are fundamental connections between the dangers of foreign oil and national identity. This is evident in descriptions of US foreign oil dependence as a threat to the American way of life, US political independence or the country's exceptional history and international responsibility. US foreign oil dependence should thus not be understood as a national security issue in the classical sense of defending state borders or ensuring the continued existence of the State. Instead, what needs to be secured against the dangers of foreign oil is the viability of a set of practices at the heart of the hydrocarbon society. These practices include, on the one hand, a specifically American way of life in which many everyday routines are directly dependent on a secure and abundant oil supply, such as driving oil-fuelled cars and purchasing products made and transported with oil. Besides such material practices, the threat posed by foreign oil dependence also concerns the viability of the various cultural discourses in which these practices have been embedded: discourses about progress, individualism, mobility, opportunity and the good life. On the other hand, foreign oil dependence also seemed to threaten the viability of well-established descriptions of the essential nature of the United States. Some of the key themes in this discourse of identity are not necessarily related to oil. For example, narratives about exceptionalism and political independence as key building blocks of the American national project go back far beyond the origins of the hydrocarbon society. Nevertheless, they also form an integral part of American national identity in this society, and foreign oil dependence has been said to undermine them both. Since the 1970s, statements about the national security threats resulting from US foreign oil dependence have been saturated with claims that dependency undermines the key tenets of the hydrocarbon society—and, indeed, threatens to destroy it. What is said to be at stake here is not the physical or even economic survival of the United States, but the survival of the United States *as a hydrocarbon society.*

This leads, finally, to the third major link between dependency and American identity. It would be wrong to see dependency only as a threat to the hydrocarbon society. The discourse of dependency actually serves to maintain, affirm and actualize that which is said to be at risk. As has been argued throughout this book, the identity of the State is relational in the sense that the identity of Self is established in relation to some Other who is described as menacing, evil or otherwise radically different. Creating state identity thus involves converting innocent difference into otherness, and the formulation of dangers in foreign policy discourse performs a vital function in this process. From this point of view, "discourses of danger" are a condition of possibility for stable national identities.[6] In other words, "The cultural production of insecurities implicates and is implicated in the cultural production of the identities of actors."[7] Debates about the continuing reliance of the nation on foreign oil, in this view, have served as an important site where the basic principles on which the country had been founded could be reaffirmed. For the past four decades, US dependence on foreign oil has afforded the nation an opportunity to affirm and renew what the United States is and what it stands for. More than anything else, this is evident in the long history of energy independence proposals, policies and declarations that have all drawn their power from the same two assumptions: that foreign oil dependence threatens the essential nature of the nation and that independence from the influence of any foreign power, including oil producers, must be the normal state of affairs for the United States. Energy independence should thus perhaps not be seen primarily as a piece of US energy policy but as a national symbol, a means of reminding Americans about the basic values and ideas on which the country was built. This goes a long way to explaining why debates about the dangers of foreign oil have been so drenched in references to American national identity, and it also helps explain why energy independence rhetoric has long remained an integral part of American political discourse, even when it seemed obvious that energy self-sufficiency was impossible and pointless.

Finally, a look ahead. What might the twenty-first century bring as far as the cultural history of dependency in the United States is concerned? From where we are standing now, it seems certain that the basic material foundation of the hydrocarbon society will persist deep into the century and perhaps beyond. Analysts predicting the long-term evolution of the US energy system tell us that oil will likely continue to provide more energy to the United States than any other source for decades to come. According to a recent prediction by the US Energy Information Administration, petroleum might still satisfy 33 percent of total US energy demand in 2040, down from 38 percent in 2012. Oil and natural gas combined will provide more than 60 percent of America's energy in 2040. While renewable energy sources might experience the highest growth rates, hydrocarbons will nevertheless continue to dominate the energy system. In the transportation sector, where oil has been particularly important, the projected picture looks similar. By 2040, renewable energy sources might account for about 6 percent of energy supply in transportation, up from 4.5 percent in 2012. Well over 90 percent of transportation fuels will be refined from fossil fuels.[8] These figures suggest that, in simple material terms, the hydrocarbon society will continue to exist and that oil will remain a raw material of central importance to economic processes, consumer choices and lifestyles in the United States for as long as we can foresee.

Predictions over such a long time frame are inherently unreliable, however, and the future might well be very different. The EIA recognizes that changes in US energy legislation, economic circumstances or exogenous events may lead to a very different development path. But energy choices are also made in the context of energy culture, and the resources offered by this culture may either work to support and maintain the present energy system or they may indeed enable new energy choices and diversions from the projected path. The future is not just shaped by physical path dependencies, market trends and technologies. Energy culture matters.

This simply means that future changes in the cultural discourses that have historically served to support the hydrocarbon society may open up possibilities for new energy choices to be made. This is true in particular as far as the representation of oil as America's lifeblood and the definition of dependency as foreign oil dependence are concerned. Both these discourses are inherently conservative in the sense that they implicitly or explicitly deny the need for an energy transition away from oil. The former claims that oil consumption is benign and natural; the latter that the key problem with oil is foreign supply, not domestic demand. As the previous chapters showed, however, both these discourses have evolved since the beginning of the twenty-first century. The lifeblood metaphor is still used today, but increasingly it has been challenged by the notion that oil is a drug to which the United States has become addicted. President Obama has spoken about the tyranny of oil over the United States.[9] These new metaphors symbolize an historical sea change in the way in which Americans have imagined their society's relationship with its most critical raw material. Reliance on oil is no longer benign, invigorating and natural but artificial, unhealthy and dangerous—something that needs to be changed. Similarly, political definitions of the problem of dependency have also changed. As Chapter 2 showed, the traditional focus on foreign oil has shifted toward a broader definition of dependency that emphasizes the dangers inherent in *all* forms of oil reliance.

The point is that such broad changes in modes of representation may ultimately affect policy choices. The emergence of the addiction metaphor and broader definitions of dependency have infused US energy debates with increasing ambiguity and uncertainty about appropriate policy solutions. As a remedy to the old problem of foreign oil dependence, energy independence has long helped maintain the hydrocarbon society. Now, however, the notion of another American energy transition comes to the fore. When Barack Obama came to power in 2009, for example, he formulated an approach to energy independence that had little to do with the goal of achieving self-sufficiency. He said: "It will be the policy of my administration to reverse our dependence on foreign oil, while building a new energy economy." Reaching energy independence, the President said, required not increased oil production but the construction of a new energy system based on efficiency, clean energy sources and reduced greenhouse gas emissions.[10] In effect, Obama equated the two notions of energy independence and energy transition. Such a definition of energy independence contains within it an existential threat to the hydrocarbon society simply because it implies an end to this energy era's material foundation: the dominance of fossil fuels. In this way, changes in American attitudes toward oil as reflected in the addiction metaphor may eventually be reflected in energy policy too, with significant effects for the energy future of the United States.

But then again, it is by no means a foregone conclusion that energy culture will continue to evolve in this direction. The ongoing US energy revolution offers a stark reminder about the power inherent in the old, time-honored discourses about energy. The energy boom is not just relevant because it brings with it sweeping changes in production quantities, energy statistics and economic prospects in the United States; it has also profoundly affected the way in which energy issues are currently being imagined in the United States. Most importantly, the US energy revolution has revived the discourse of abundance. As two energy experts argued in a recent *New York Times* editorial, until recently the world of energy was shaped by "the axiom of 'resource scarcity'". Thanks to the US energy boom, they observed, "we have ushered in an era in which 'resource abundance' will be the norm".[11] This return of material abundance raises an interesting question. Will it also reactivate the cultural discourse of abundance that has been so important in shaping definitions of dependency historically? Is it conceivable that the pendulum will swing back from a reading of dependency based more and more on the addiction metaphor to one in which oil, and increasingly natural gas, will again be conceptualized as America's lifeblood? To some extent, this is what has happened over the past few years. As the second chapter argued, the current US energy boom is frequently represented in terms of its positive effects on the issue of US foreign oil dependence and the possibility of finally achieving energy independence. This indicates the possibility of a return to a more traditional reading of the problem of dependency because it marginalizes the alternative view of depicting oil reliance as an addiction that the country needs to overcome as quickly as possible. Even Barack Obama, who started his presidency by speaking about the tyranny of oil in the United States and the urgent need to build a clean energy economy, began to represent increased production of fossil fuels as an achievement of his administration: "When it comes to oil production, under my administration, America is producing more oil today than at any time in the last 8 years. That is a fact." This was good news, according to the President, since it meant reduced dependence on foreign oil: "Since I took office, America's dependence on foreign oil has gone down every single year—every single year. In fact, in 2010, it was under 50 percent for the first time in 13 years—for the first time."[12] There has been a certain tendency to make sense of the current developments in the field of energy in terms of well-established descriptions of the role of oil in American society. Within this discourse, the re-emerging American energy abundance indeed appears as a positive achievement, even a breakthrough.

Whether this represents merely a temporary shift or a lasting return to more traditional definitions of dependency remains to be seen. There are good reasons to be skeptical about the possibility that the discourse of abundance will return in the form it took during the first half of the twentieth century—with its focus on limitlessness, infinite economic growth and material progress. Environmental concerns about the adverse effects of burning fossil fuels appear very strong now. Even if oil and gas supply may be imagined in terms of limitlessness, it would be hard to reconcile this with the increasingly widespread sentiment that the world's ecosystems are finite and limited and that we are currently treading dangerously close to those limits. It is perhaps unrealistic to expect that oil discourse will evolve very strongly in either direction. The ambiguity that has crept into the metaphorical representation of oil as well as definitions of the problem of dependency might well continue to shape energy debates for the foreseeable future. What we are witnessing today, in effect, is a fundamental cultural and discursive struggle in the United States over the major energy choices that will shape how we will live in the twenty-first century. Providing insights and resources in order to make sense of how we got here has been a central ambition of this book. For me, the most beautiful aspect of US energy culture is that its evolution is uncertain, unpredictable and somewhat mysterious. The best we can do is to observe carefully the subtle twists and turns in American discourses about energy, and try to come to a deeper understanding of what these changes may signify for the direction of energy and foreign policy and for American society at large. At the same time, it is worth remembering that energy discourse is created, kept alive and altered through the practices of actual individuals. Unlikely as it may be, if you are among those whose voices are heard in the United States, you can always try to give things a gentle nudge in the desired direction.

NOTES

Introduction A Culture of Dependency

1. Quoted in Jad Mouawad and Matthew L. Wald, "The oil uproar that isn't", *New York Times*, 12 July 2005.
2. Richard M. Nixon, "Special message to the Congress on the energy crisis", 23 January 1974, in Gerhard Peters and John T. Woolley (eds), *The American Presidency Project*, *Public Papers of the Presidents*. Available at www.presidency.ucsb.edu/ws/?pid=4241 (accessed 31 October 2013).
3. Figure for 2013. These statistics are taken from Table 3.3a in US Energy Information Administration, *Monthly Energy Review: May 2014*, DOE/EIA-0035 (2014), p. 41. Available at www.eia.gov/mer (accessed 7 June 2014).
4. Jerry Taylor and Peter Van Doren, "The energy security obsession", *Georgetown Journal of Law and Public Policy* 6/2 (2008), pp. 475, 486.
5. Robert Bryce, *Gusher of Lies: The Dangerous Delusions of 'Energy Independence'* (New York, 2008), p. 7.
6. US Senate Committee on Foreign Relations, *Hearing on Energy Security and Oil Dependence*, 109th Cong., 2nd sess., 16 May 2006, pp. 14, 21–2 (statement of Jason S. Grumet, executive director of the National Commission on Energy Policy).
7. Leonardo Maugeri, *The Age of Oil: The Mythology, History, and Future of the World's Most Controversial Resource* (Westport, CT, 2006), pp. xi, 267.
8. Gal Luft and Anne Korin, "The myth of US energy dependence", *Foreign Affairs*, 15 October 2013. Available at www.foreignaffairs.com/articles/140172/gal-luft-and-anne-korin/the-myth-of-us-energy-dependence (accessed 1 November 2013).
9. For instance, John S. Duffield, *Over a Barrel: The Costs of US Foreign Oil Dependence* (Stanford, CA, 2008); Ian W. H. Parry and Joel Darmstadter, "The costs of US oil dependency", *Resources for the Future,* Discussion Paper 03–59 (2003). Available at www.rff.org/documents/rff-dp-03–59.pdf (accessed 16 July 2014); James L. Plummer (ed.), *Energy Vulnerability* (Cambridge, MA, 1982).

For a good introduction to this topic, also see Douglas R. Bohi, Michael A. Toman and Margaret A. Walls, *The Economics of Energy Security* (Boston, MA, 1996), pp. 59–71.

10. For "myths and realities" see Leonardo Maugeri, "Not in oil's name", *Foreign Affairs* 82/4 (2003), p. 165. For "rhetoric and reality" see Bryce, *Gusher of Lies*, p. 7.
11. Besides the literature already quoted, a good example is Keith Crane et al., *Imported Oil and US National Security* (Santa Monica, CA, 2009). This study is described by the authors as a critical evaluation of commonly suggested links between imported oil and national security with the explicit aim of enabling better policy making.
12. A Washington-based energy analyst in a personal conversation with the author, 10 November 2011.
13. Maugeri, *The Age of Oil*, p. 260.
14. Amitav Ghosh, "Petrofiction: the oil encounter and the novel", *New Republic* (2 March 1992), p. 30.
15. Robert W. Cox, "Social forces, states and world orders: beyond international relations theory", *Millennium: Journal of International Studies* 10/2 (1986), pp. 128–9.
16. This approach is inspired by, embedded in and an advancement of a diverse literature, which, very broadly speaking, emphasizes that energy issues are social constructions. Melani McAlister, Martin Melosi, David Nye, Robert Vitalis, Natasha Zaretsky and others have written important studies from this perspective, works which will be referenced extensively throughout this book.
17. Due in large part to increased domestic production, US oil import reliance seems to be in secular decline now. The US Energy Information Administration currently assumes that by the year 2040 US dependence on foreign oil may be somewhere in the neighborhood of 32 percent of domestic demand. If certain economic and political conditions are in place, however, the country might even become a net exporter of oil. These predictions are based on the different scenarios in US Energy Information Administration, *Annual Energy Outlook 2014*, DOE/EIA-0383 (2014), pp. ES-2, ES-3. Available at www.eia.gov/forecasts/aeo (accessed 9 June 2013).
18. Cox, "Social forces, states and world orders", p. 129.
19. The theoretical basis for this claim is the notion that objects in the world are constituted as meaningful and significant only within discourse. According to Epstein's definition, "A *discourse* is a cohesive ensemble of ideas, concepts, and categorizations about a specific object that frame that object in a certain way and, therefore, delimit the possibilities for action in relation to it." Charlotte Epstein, *The Power of Words in International Relations: Birth of an Anti-Whaling Discourse* (Cambridge, MA, 2008), p. 2. Emphasis in the original. It should be noted, however, that many different approaches rely on a discursive epistemology. This book adopts what is sometimes called a "critical constructivist" perspective. A good summary of this position can be found in Jutta Weldes et al., "Introduction: constructing insecurity", in Jutta Weldes et al. (eds), *Cultures of*

Insecurity: States, Communities, and the Production of Danger (Minneapolis, MN, 1999).

0. Michael T. Klare, *Blood and Oil: The Dangers and Consequences of America's Growing Petroleum Dependency* (New York, 2004), p. xiii. Emphasis in the original.

1. Ian Rutledge, *Addicted to Oil: America's Relentless Drive for Energy Security* (London, 2006), p. 78.

2. Bruce A. Beaubouef, *The Strategic Petroleum Reserve: US Energy Security and Oil Politics, 1975–2005* (College Station, TX, 2007), p. 220.

3. This proposition is based on the view that identity is performative and relational. Identity is performative because it is not given or inherent in the subject to which it belongs but has to be constantly made and remade. It is relational because it is always established in relation to difference, to what it is not. As William Connolly explains, "An identity is established in relation to a series of differences that have become socially recognized. These differences are essential to its being. If they did not coexist as differences, it would not exist in its distinctness and solidity." William E. Connolly, *Identity/Difference: Democratic Negotiations of Political Paradox* (Ithaca, NY, 1991), p. 64.

4. Foreign policy can be conceptualized as a meaning-making activity. As David Campbell puts it, foreign policy is part of a "discourse of danger". Within such discourses, all kinds of dangers and threats to the state are formulated by policy makers and other members of the policy community, with the ultimate effect of reproducing the identity of the state. David Campbell, *Writing Security: United States Foreign Policy and the Politics of Identity* (Manchester, 1992), pp. 12, 53–6.

5. Jutta Weldes argues that the authority to define foreign policy issues is unevenly distributed among different actors in what she calls the US "security imaginary", defined as a pre-existing "structure of well-established meanings and social relations out of which representations of the world of international relations are created". High-ranking representatives of the state naturally play a central role in this process. Other relevant speakers include the "ideological state apparatus", which contains cultural and educational institutions and the news media, and the "intellectual state apparatus" whose members derive their authority from claims to expert knowledge: academic institutions, think tanks and the like. Jutta Weldes, *Constructing National Interests: The United States and the Cuban Missile Crisis* (Minneapolis, MN, 1999), pp. 10, 108. Also see Lene Hansen, *Security as Practice: Discourse Analysis and the Bosnian War* (London, 2006), p. 66.

Chapter 1 Constructing the Hydrocarbon Society

1. Curtis Kenyon, *Tulsa*, motion picture by Walter Wanger Productions (1949). Available at www.youtube.com/watch?v=WO7WPxZt63w (accessed 16 July 2014).

2. Robert P. Russell, "Oil: molecular treasure chest", originally published in the *Kiwanis* magazine, December 1945, Standard Oil of New Jersey reprint, in *ExxonMobil Historical Collection*, Dolph Briscoe Center for American History, University of Texas at Austin, part 6, box 2.207/L13A.
3. Daniel Yergin, *The Prize: The Epic Quest for Oil, Money & Power* (New York, 2008), p. xvi.
4. In 2012, the *Journal of American Studies* devoted a special issue to the question of how oil has been imagined in different cultural texts. See Ross Barrett and Daniel Worden, "Oil culture: guest editors' introduction", *Journal of American Studies* 46/2 (2012), pp. 269–72.
5. Martin Melosi, "Energy transitions in historical perspective", in Brendan M. Dooley (ed.), *Energy and Culture: Perspectives on the Power to Work* (Aldershot, 2006), p. 3.
6. Figures for 1860–1940 calculated based on data in Sam H. Schurr and Bruce C. Netschert, *Energy in the American Economy, 1850–1975* (Baltimore, MD, 1960), pp. 36, 145. Figures for 1950 and later based on Table 1.3 in US Energy Information Administration, *Monthly Energy Review: May 2014*, p. 7.
7. David E. Nye, *Consuming Power: A Social History of American Energies* (Cambridge, MA, 2001), p. 8.
8. David E. Nye, "Energy narratives", *American Studies in Scandinavia* 25/2 (1993), pp. 73–91.
9. Nye, *Consuming Power*, p. 4.
10. Ibid., p. 3. The term "technological momentum" was coined by Thomas Hughes in Thomas P. Hughes, "Technological momentum", in Merrit Roe Smith and Leo Marx (eds), *Does Technology Drive History?: The Dilemma of Technological Determinism* (Cambridge, MA, 1994).
11. See Campbell, *Writing Security*, p. 4; Roxanne Lynn Doty, "Foreign policy as social construction: a post-positivist analysis of US counterinsurgency policy in the Philippines", *International Studies Quarterly* 37/3 (1993), p. 298.
12. Campbell, *Writing Security*, p 6.
13. David S. Painter, *Oil and the American Century: The Political Economy of US Foreign Oil Policy, 1941–1954* (Baltimore, MD, 1986), p. 209.
14. Melani McAlister, *Epic Encounters: Culture, Media, and US Interests in the Middle East, 1945–2000* (Berkeley, CA, 2001), p. 34. A great study of frontier discourse in Saudi Arabia is Robert Vitalis, *America's Kingdom: Mythmaking on the Saudi Oil Frontier* (Stanford, CA, 2007).
15. Willard Van Dyke, *American Frontier*, short film by Affiliated Film Producers and the American Petroleum Institute (1953). Available at www.archive.org/details/American1953 (accessed 16 July 2014).
16. Roger M. Olien and Diana Davids Olien, *Oil and Ideology: The Cultural Creation of the American Petroleum Industry* (Chapel Hill, NC, 2000), pp. 42, 251.
17. A good example of these views is provided by the public outrage that accompanied the energy crises of the 1970s. See, for instance, Yergin, *The Prize*, pp. 638–46; Nye, *Consuming Power*, pp. 222, 229–34.

18. Mobil Oil Company, Memo from Herbert Schmertz to List B, "Public affairs advertising", 19 January 1972, in *ExxonMobil Historical Collection*, Dolph Briscoe Center for American History, University of Texas at Austin, part 2, box 2.207/E85. The collection also includes reprints of the "Op-Ed" series.
19. This is a finding emerging from a series of Gallup polls covering the period from 2001 to 2012. See Gallup, "Americans rate computer industry best, oil and gas worst", 16 August 2012. Available at www.gallup.com/poll/156713/Americans-Rate-Computer-Industry-Best-Oil-Gas-Worst.aspx (accessed 8 November 2013).
20. George G. MacCurdy, *Human Origins: A Manual of Prehistory*, vol. II (New York, 1933), pp. 134–5.
21. Bryce, *Gusher of Lies*, p. 19.
22. Yergin, *The Prize*, p. 653. Today, economists point out that due to saturation effects there is no simple linear relationship between those variables. José Goldemberg and Oswaldo Lucon, *Energy, Environment and Development* (London, 2010), pp. 65–7.
23. Stefania Barca, "Energy, property, and the Industrial Revolution narrative", *Ecological Economics* 70/7 (2011), pp. 1,309–10.
24. Yergin, *The Prize*, p. xvi.
25. Amory B. Lovins, "Energy strategy: the road not taken?", *Foreign Affairs* 55/1 (1976), p. 66.
26. Nye, *Consuming Power*, p. 217.
27. Martin Melosi, *Coping with Abundance: Energy and Environment in Industrial America* (Philadelphia, PA, 1985), p. 3.
28. Nye, *Consuming Power*, p. 217. Also see Nye, "Energy narratives", pp.77–9.
29. David M. Potter, *People of Plenty: Economic Abundance and the American Character* (Chicago, 1954), p. 84.
30. Eugene Holman, "Our inexhaustible resources", originally published in the *Atlantic Monthly*, June 1952, Standard Oil of New Jersey reprint, in *ExxonMobil Historical Collection*, Dolph Briscoe Center for American History, University of Texas at Austin, part 6, box 2.207/L13A. Cf. Timothy Mitchell, *Carbon Democracy: Political Power in the Age of Oil* (London, 2011), p. 188.
31. Melosi, "Energy transitions in historical perspective", p. 7.
32. Nye, *Consuming Power*, pp. 187, 202.
33. Standard Oil Company (New Jersey), "Petroleum in the world", booklet for educational use, 1945, in *ExxonMobil Historical Collection*, Dolph Briscoe Center for American History, University of Texas at Austin, part 6, box 2.207/L13A, pp. 5, 11, 29–30.
34. Mitchell, *Carbon Democracy*, p. 139.
35. See Nye, *Consuming Power*, ch. 6.
36. Lizabeth Cohen, *A Consumers' Republic: The Politics of Mass Consumption in Postwar America* (New York, 2003), p. 119.
37. Nye, *Consuming Power*, p. 169.

38. Image in Esso Standard Oil Company, "What makes a car go", children's book, 1957, in *ExxonMobil Historical Collection*, Dolph Briscoe Center for American History, University of Texas at Austin, part 6, box 2.207/L13A. Reproduced courtesy of ExxonMobil Corporation.
39. This campaign is described in David E. Nye, *Image Worlds: Corporate Identities at General Electric, 1890–1930* (Cambridge, MA, 1985), pp. 124–32.
40. Mark Gottdiener, "Approaches to consumption: classical and contemporary perspectives", in Mark Gottdiener (ed.), *New Forms of Consumption: Consumers, Culture, and Commodification* (Lanham, MD, 2000), pp. 21–5.
41. NSC 68, "United States objectives and programs for national security", 14 April 1950. Reprinted in US Department of State, *Foreign Relations of the United States (FRUS): 1950, Volume I (National security affairs, foreign economic policy)* (Washington, DC, 1977), pp. 235–92 (quotes on p. 239).
42. See Mike Featherstone, *Consumer Culture and Postmodernism* (Los Angeles, 2007), pp. xv, xix; Cohen, *A Consumers' Republic*, pp. 124–6.
43. David H. Fischer, *Liberty and Freedom: A Visual History of America's Founding Ideas* (Oxford, 2005), pp. 669–70.
44. See ibid., p. 670.
45. Eugene Holman, "Oil – product and pillar of peace", speech delivered before the Chamber of Commerce of the State of New York, 6 April 1944, in *ExxonMobil Historical Collection*, Dolph Briscoe Center for American History, University of Texas at Austin, part 6, box 2.207/L13C.
46. Carl Urbano, *Destination Earth*, animated short film by Sutherland Productions and the American Petroleum Institute (1956). Available at http://archive.org/details/Destinat1956 (accessed 16 July 2014).
47. Melosi, *Coping with Abundance*, p. 104.
48. John Urry, "The 'system' of automobility", *Theory, Culture & Society* 21/4 (2004), pp. 25–6.
49. Melosi, *Coping with Abundance*, p. 269.
50. Rutledge, *Addicted to Oil*, p. 19.
51. A former presidential science advisor. Quoted in ibid., p. 14.
52. For an example of this position see, Edwin Black, *Internal Combustion: How Corporations and Governments Addicted the World to Oil and Derailed the Alternatives* (New York, 2006). On this debate, see David J. St. Clair, *The Motorization of American Cities* (New York, 1986).
53. Bruce Barton, *The Man Nobody Knows* (Indianapolis, IN, 1925), p. 140.
54. This campaign, and other corporate efforts to advertise gasoline, are described in detail in Yergin, *The Prize*, pp. 191–4, 530–2.
55. Quoted in Cohen, *A Consumers' Republic*, p. 293.
56. Potter, *People of Plenty*, pp. 91–3.
57. Quoted in ibid., p. 96.
58. Institute of Labor and Industrial Relations, *Contribution of the Automotive Industry to the US Economy in 1998*, a study prepared for the Alliance of Automobile Manufacturers and the Association of International Automobile Manufacturers

(2001), p. 49. Available at http://deepblue.lib.umich.edu/bitstream/2027.42/1483/2/94310.0001.001.pdf (accessed 8 November 2013).
9. Melosi, *Coping with Abundance*, p. 105.
0. Nye, *Consuming Power*, p. 176.
1. Rutledge, *Addicted to Oil*, p. 19.
2. Nye, *Consuming Power*, p. 206.
3. Melosi, *Coping with Abundance*, p. 110.
4. Ibid., pp. 109–11, 270.
5. Nye, *Consuming Power*, p. 206.
6. Ibid., p. 180.
7. Melosi, *Coping with Abundance*, p. 112.
8. Nye, *Consuming Power*, p. 177.
9. Yergin, *The Prize*, p. 534.
0. US Senate Committee on Finance, *Bills to Encourage National Industrial Recovery, to Foster Fair Competition, and to Provide for the Construction of Certain Useful Public Works, and for Other Purposes, Hearings on S. 1712 and H.R. 5755*, 73rd Cong., 1st sess., 18–20 May 1933, p. 42 (statement of Harold L. Ickes, secretary of the interior).
1. Securing America's Future Energy (SAFE), *Oil Dependence: A Threat to US Economic & National Security*, Policy Report (2006). Available at www.securenergy.org/sites/default/files/155_Briefing-OilDependence.pdf (accessed 8 November 2013).
2. On the role of metaphors in shaping cognitive processes, see Paul Chilton and George Lakoff, "Foreign policy by metaphor", in Christina Schäffner and Anita Wenden (eds), *Language and Peace* (Aldershot, 1995); George Lakoff and Mark Johnson, *Metaphors We Live By* (Chicago, 1980). On the history of body-politic metaphors more specifically, see Arnold D. Harvey, *Body Politic: Political Metaphor and Political Violence* (Newcastle-upon-Tyne, 2007).
3. Upton Sinclair, *Oil!* (London, 2008), p. 5.
4. Rutledge, *Addicted to Oil*, p. 20.

Chapter 2 Defining Dependency

1. George W. Bush, "Address before a joint session of the Congress on the State of the Union", 31 January 2006, in Peters and Woolley, *The American Presidency Project*. Available at www.presidency.ucsb.edu/ws/?pid=65090 (accessed 31 October 2013).
2. "Bush's goals on energy quickly find obstacles", *New York Times*, 2 February 2006.
3. Thomas L. Friedman, "Will pigs fly?", *New York Times*, 3 February 2006.
4. Barton, *The Man Nobody Knows*, p. 140.
5. Cartoon by Kevin Kallaugher, *The Economist*, 17 June 2010. Reproduced with kind permission of Kevin KAL Kallaugher, www.kaltoons.com.
6. Michel Foucault, *Discipline and Punish: The Birth of the Prison* (London, 1977), p. 31.

7. Jens Bartelson, *A Genealogy of Sovereignty* (Cambridge, 1995), p. 8.
8. Ernest H. Davenport and Sidney R. Cooke, *The Oil Trusts and Anglo-American Relations* (New York, 1924), p. 67.
9. Harold L. Ickes, "After the deluge: what price gasoline?", *Saturday Evening Post* (16 February 1935), p. 6.
10. Yergin, *The Prize*, p. 230.
11. Melosi, *Coping with Abundance*, p. 152.
12. Ickes, "After the deluge", pp. 6, 42.
13. See Olien and Davids Olien, *Oil and Ideology*, p. 120.
14. See the discussion in Mitchell, *Carbon Democracy*, pp. 39–42, quote on p. 41.
15. For an overview of the policy debate and the pro-rationing system, see Yergin, *The Prize*, ch. 13; Olien and Davids Olien, *Oil and Ideology*, ch. 8; Richard H. K. Vietor, *Energy Policy in America since 1945: A Study of Business-Government Relations* (Cambridge, 1984), ch. 2.
16. Vietor, *Energy Policy in America*, p. 121.
17. Dwight D. Eisenhower, "Annual message to the Congress on the State of the Union", 10 January 1957, in Peters and Woolley, *The American Presidency Project*. Available at www.presidency.ucsb.edu/ws/?pid=11029 (accessed 18 November 2013).
18. See Mitchell, *Carbon Democracy*, pp. 177–81. Also see Yergin, *The Prize*, p. 572.
19. This debate was influenced by two factors. One was a national security dilemma: cheap foreign oil was available in abundance to help conserve domestic reserves for the future, but such long-term supply security came at the risk of more immediate foreign oil dependence. Secondly, there was an important division within the American oil industry, whereby independent producers feared that cheap foreign oil would drive them out of business while the international majors promoted imports of their oil. All sides argued on grounds of national security. A detailed account of these debates and the quota system can be found in Vietor, *Energy Policy in America*, chs 5–6.
20. Ibid., p. 115.
21. Data from table 3.1 in US Energy Information Administration, *Monthly Energy Review: May 2014*, p. 37.
22. US Department of State, "The international oil industry through 1980", December 1971. Quoted in Yergin, *The Prize*, p. 572.
23. Iraq, the tenth member of OAPEC, did not participate in the decision but later implemented the destination embargo against the United States. Formally, the decision was taken in the framework of the Conference of Arab Oil Ministers. See Ian Skeet, *OPEC: Twenty-Five Years of Prices and Politics* (Cambridge, 1988), p. 100.
24. This account of the 1973 and 1979 oil shocks is based on Fiona Venn, *The Oil Crisis* (London, 2002), ch. 1.
25. There is no consensus on the exact relationship between oil and the economy, however. For a discussion of the economics behind this assumption, see Robert B. Barsky and Lutz Kilian, "Oil and the macroeconomy since the 1970s",

Journal of Economic Perspectives 18/4 (2004), pp. 115–34. The claim that the oil shock was a primary cause of the economic troubles affecting the global economy in the early 1970s has been criticized from different points of view. Cf. Mitchell, *Carbon Democracy*, ch. 7; Venn, *Oil Crisis*, ch. 5; Simon Bromley, *American Hegemony and World Oil: The Industry, the State System and the World Economy* (University Park, PA, 1991), ch. 4.

6. November 1973 and March 1975 represent the peak and trough of the US business cycle according to data in The National Bureau of Economic Research (NBER), *US Business Cycle Expansions and Contractions* (2013). Available at www.nber.org/cycles/cyclesmain.html (accessed 16 July 2014).
7. Beaubouef, *Strategic Petroleum Reserve*, pp. 22–3, quote on p. 22. Like many others, however, Beaubouef also argues that major government interventions into the US economy in the form of wage and price controls, as well as gasoline rationing, significantly exacerbated the problems.
8. Yergin, *The Prize*, p. 592. Cf. Nye, *Consuming Power*, ch. 8.
9. Beaubouef, *Strategic Petroleum Reserve*, pp. 17–18. For a detailed discussion of the energy crisis as a social crisis with respect to American middle-class families, see Natasha Zaretsky, *No Direction Home: The American Family and the Fear of National Decline, 1968–1980* (Chapel Hill, NC, 2007), ch. 2.
0. Karen R. Merrill, *The Oil Crisis of 1973–1974: A Brief History with Documents* (Boston, MA, 2007), p. 4.
1. Rutledge, *Addicted to Oil*, p. 46.
2. Yergin, *The Prize*, p. 598.
3. James E. Akins, "The oil crisis: this time the wolf is here", *Foreign Affairs* 51/3 (1973), pp. 462–90.
4. Melosi, *Coping with Abundance*, p. 277.
5. Stuart Hall, "Signification, representation, ideology: Althusser and the post-structuralist debates", *Critical Studies in Mass Communication* 2/2 (1985), p. 113, fn. 2.
6. Alexander Wendt, *Social Theory of International Politics* (Cambridge, 1999), p. 76. Cf. Peter L. Berger and Thomas Luckmann, *The Social Construction of Reality: A Treatise in the Sociology of Knowledge* (Garden City, NY, 1966), p. 89.
7. Richard M. Nixon, "Special message to the Congress on energy resources", 4 June 1971, in Peters and Woolley, *The American Presidency Project*. Available at www.presidency.ucsb.edu/ws/?pid=3038 (accessed 22 November 2013).
8. Richard M. Nixon, "Address on the State of the Union delivered before a joint session of the Congress", 30 January 1974, in ibid. Available at www.presidency.ucsb.edu/ws/?pid=4327 (accessed 22 November 2013).
9. Nixon, "Special message to the Congress on energy resources", 4 June 1971.
0. Merrill, *Oil Crisis*, p. 22.
1. "Nation's energy crisis: Is unbridled growth indispensable to the good life?", *New York Times*, 8 July 1971.
2. Donella Meadows et al., *The Limits to Growth: A Report for the Club of Rome's Project on the Predicament of Mankind* (London, 1972).

43. Paul R. Ehrlich, *The Population Bomb* (New York, 1971); Ernst F. Schumacher *Small is Beautiful: A Study of Economics as if People Mattered* (London, 1973).
44. Lovins, "Energy strategy: the road not taken?", pp. 65–96.
45. See Daniel Horowitz, *The Anxieties of Affluence: Critiques of American Consume Culture, 1939–1979* (Amherst, NY, 2004).
46. See Mitchell, *Carbon Democracy*, pp. 188–90.
47. Melosi, *Coping with Abundance*, p. 266. A good overview of the environmenta issues raised by oil can be found in chs 14–16 of that book.
48. Caroll L. Wilson, "A plan for energy independence", *Foreign Affairs* 51/4 (1973), p. 657.
49. "Will a proud Mexico share its oil wealth?", *Chicago Tribune*, 20 May 1978.
50. Zaretsky, *No Direction Home*, quote on p. 80.
51. See, for example, Lovins, "Energy strategy: the road not taken?", pp. 65–96 Daniel Yergin, "US energy policy: transition to what?", *World Today* 35/3 (1979), pp. 81–91; Lewis J. Perelman, "Speculations on the transition to sustainable energy", *Ethics* 90/3 (1980), pp. 392–416. Also see the discussion in Melosi, *Coping with Abundance*, pp. 312–19.
52. Gerald R. Ford, "Address to a joint session of the Congress on the economy" 8 October 1974, in Peters and Woolley, *The American Presidency Project*. Available at www.presidency.ucsb.edu/ws/?pid=4434 (accessed 26 November 2013).
53. James E. Carter, "Address to the nation on energy", 18 April 1977, in ibid Available at www.presidency.ucsb.edu/ws/?pid=7369 (accessed 31 October 2013); James E. Carter, "Address to the nation on energy and national goals 'the Malaise Speech'", 15 July 1979, in ibid. Available at www.presidency.ucsb.edu/ws/?pid=32596 (accessed 26 November 2013).
54. Horowitz, *The Anxieties of Affluence*, p. 206.
55. James E. Carter, "The President's news conference", 20 July 1978, in Peters and Woolley, *The American Presidency Project*. Available at www.presidency.ucsb.edu/ws/?pid=31107 (accessed 26 November 2013).
56. Francisco Parra, *Oil Politics: A Modern History of Petroleum* (London, 2010) p. 254.
57. Carter, "Address to the nation on energy", 18 April 1977.
58. The events leading up to the OAPEC oil embargo are well documented. See, for example, Yergin, *The Prize*, pp. 584–91; Venn, *Oil Crisis*, pp. 9–15.
59. Nixon, "Special message to the Congress on the energy crisis", 23 January 1974.
60. Richard M. Nixon, "Radio address about the national energy crisis", 19 January 1974, in Peters and Woolley, *The American Presidency Project*. Available at www.presidency.ucsb.edu/ws/?pid=4208 (accessed 27 November 2013).
61. Gerald R. Ford, "Special message to the Congress on energy", 7 January 1977, in ibid. Available at www.presidency.ucsb.edu/ws/?pid=5542 (accessed 17 December 2013).
62. James E. Carter, "The President's news conference", 28 July 1977, in ibid. Available at www.presidency.ucsb.edu/ws/?pid=7878 (accessed 27 November 2013).

63. Carter, "Address to the nation on energy", 18 April 1977.
64. Stephen J. Randall, *United States Foreign Oil Policy Since World War I: For Profits and Security* (Montreal, 2005), p. 15.
65. Although the report concluded that a liberalization of oil imports would not significantly undermine US national security. See US Cabinet Task Force on Oil Import Control, *The Oil Import Question: A Report on the Relationship of Oil Imports to the National Security* (Washington, DC, 1970).
66. Joseph Alsop, "Oil blackmail threatens US independence", *Washington Post*, 21 November 1973.
67. Poll conducted by Louis Harris & Associates, 27 December 1977 through 10 January 1978. Data provided courtesy of the Roper Center for Public Opinion Research, University of Connecticut.
68. Poll conducted by Opinion Research Corporation, 28 January 1978 through 8 February 1978. Data provided courtesy of the Roper Center for Public Opinion Research, University of Connecticut.
69. Question asked in a series of polls conducted by Roper, 1977–89. Quoted in Barbara C. Farhar, "Public opinion about energy", *Public Opinion Quarterly* 58/4 (1994), pp. 610, 627.
70. Nye, "Energy narratives", pp. 77–9.
71. Nye, *Consuming Power*, p. 217.
72. Richard M. Nixon, "Address to the nation about policies to deal with the energy shortages", 7 November 1973, in Peters and Woolley, *The American Presidency Project*. Available at www.presidency.ucsb.edu/ws/?pid=4034 (accessed 18 September 2012).
73. Yergin, *The Prize*, pp. 599–601.
74. Parra, *Oil Politics*, p. 186.
75. US Federal Energy Administration, *Project Independence Report* (Washington, DC, 1974). The report argued that rather than achieving self-sufficiency, US energy policy should focus on reducing the vulnerability resulting from import reliance—for example, through strategic stockpiling. See Beaubouef, *Strategic Petroleum Reserve*, pp. 19–22.
76. For example, the deadline for achieving energy independence was moved back to 1985. Also, Ford defined energy independence not as complete self-sufficiency but as invulnerability to foreign disruption of oil. See Gerald R. Ford, "Address to the nation on energy and economic programs", 13 January 1975, in Peters and Woolley, *The American Presidency Project*. Available at www.presidency.ucsb.edu/ws/?pid=4916 (accessed 29 November 2013); Gerald R. Ford, "Address before a joint session of the Congress reporting on the State of the Union", 15 January 1975, in ibid. Available at www.presidency.ucsb.edu/ws/?pid=4938 (accessed 29 November 2013).
77. "Energy freedom" is Carter's term. James E. Carter, "The President's news conference", 28 November 1979, in ibid. Available at www. presidency.ucsb.edu/ws/?pid=31752 (accessed 29 July 2014). The idea of "energy non-dependence" appears in Secretary of Energy Hodel's 1983 National Energy

Plan. Donald P. Hodel, "National Energy Policy Plan: a preview", *Vital Speeches of the Day* 50/5 (1983), pp. 130–3.

78. Richard M. Nixon, "Address to the nation about national energy policy", 25 November 1973, in Peters and Woolley, *The American Presidency Project*. Available at www.presidency.ucsb.edu/ws/?pid=4051 (accessed 29 November 2013).
79. Stefan A. Halper and Jonathan Clarke, *The Silence of the Rational Center: Why American Foreign Policy is Failing* (New York, 2007), pp. 4–5.
80. These statistics are taken from tables 1.3 and 3.3a in US Energy Information Administration, *Monthly Energy Review: May 2014*, pp. 7, 41.
81. Mobil Oil Company, "The national guilt complex. Achievement or original sin. Production vs. atonement", advertisement published in various magazines, April 1978, in *ExxonMobil Historical Collection*, Dolph Briscoe Center for American History, University of Texas at Austin, part 2, box 2.207/E85.
82. Mobil Oil Company, "Should US energy policy make me feel guilty about my lifestyle?", advertisement published in the *New York Times*, 25 August 1977, in *ExxonMobil Historical Collection*, Dolph Briscoe Center for American History, University of Texas at Austin, part 2, box 2.207/E85.
83. Ronald W. Reagan, "Cleveland Speech", 10 September 1980. Quoted in Nye, *Consuming Power*, p. 235.
84. Ibid., pp. 236–7.
85. Barbara Rudolph, "Step on the gas, pay the price", *Time*, 3 April 1989.
86. This paragraph draws on Yergin's account of the emergence of climate change as a political issue in the United States, in Daniel Yergin, *The Quest: Energy, Security, and the Remaking of the Modern World* (New York, 2011), ch. 23. Also see Cathleen Fogel, "Constructing progressive climate change norms: the US in the early 2000s", in Mary E. Pettenger (ed.), *The Social Construction of Climate Change: Power, Knowledge, Norms, Discourses* (Aldershot, 2007).
87. A good summary of Hubbert's work is to be found in David Strahan, *The Last Oil Shock: A Survival Guide to the Imminent Extinction of Petroleum Man* (London, 2008), ch. 2.
88. Colin J. Campbell and Jean H. Laherrère, "The end of cheap oil", *Scientific American Magazine* (March 1998), pp. 78–83.
89. There have been a number of books on the topic written for a general audience. See, for example, David L. Goodstein, *Out of Gas: The End of the Age of Oil* (New York, 2004); Paul Roberts, *The End of Oil: On the Edge of a Perilous New World* (Boston, 2005); and Richard Heinberg, *The Party's Over: Oil, War and the Fate of Industrial Societies* (Gabriola Island, BC, 2005).
90. Imre Szeman, "System failure: oil, futurity, and the anticipation of disaster", *South Atlantic Quarterly* 106/4 (2007), pp. 805–23.
91. Albert A. Gore, "Still hooked on oil", *New York Times*, 6 March 1991.
92. Charles Krauthammer, "A nation of oil addicts", *Washington Post*, 17 March 2000.

93. US Senate Committee on Foreign Relations, *Hearing on Energy Security and Oil Dependence*, p. 1 (statement of Richard G. Lugar, chairman of the US Senate Committee on Foreign Relations).
94. Ibid., pp. 22–3 (statement of Jason S. Grumet, executive director of the National Commission on Energy Policy).
95. Barack H. Obama, "Remarks at the Department of Energy", 5 February 2009, in Peters and Woolley, *The American Presidency Project*. Available at www.presidency.ucsb.edu/ws/?pid=85735 (accessed 5 December 2013).
96. Barack H. Obama, "Remarks on energy", 26 January 2009, in ibid. Available at www.presidency.ucsb.edu/ws/?pid=85689 (accessed 4 December 2013).
97. See Stanley Reed and Alison Fitzgerald, *In Too Deep: BP and the Drilling Race That Took It Down* (Hoboken, NJ, 2011), ch. 5.
98. Barack H. Obama, "Address to the nation on the oil spill in the Gulf of Mexico", 15 June 2010, in Peters and Woolley, *The American Presidency Project*. Available at www.presidency.ucsb.edu/ws/?pid=88033 (accessed 4 December 2013).
99. See, for example, Steve Hallett and John Wright, *Life Without Oil: Why We Must Shift to a New Energy Future* (Amherst, NY, 2011); Geert Verbong and Derk Loorbach, *Governing the Energy Transition: Reality, Illusion or Necessity?* (New York, 2012).
100. Donald Kennedy, "Beyond petroleum?", *Science* 329/5993 (2010), p. 727.
101. Thomas L. Friedman, "No fooling mother nature", *New York Times*, 4 May 2010.
102. Yergin, *The Prize*, p. 641.
103. Toby Bolsen and Fay Lomax Cook, "Public opinion on energy policy, 1974–2006", *Public Opinion Quarterly* 72/2 (2008), pp. 364–88.
104. Ronald W. Reagan, "Message to the Congress transmitting the National Energy Policy Plan", 17 July 1981, in Peters and Woolley, *The American Presidency Project*. Available at www.presidency.ucsb.edu/ws/?pid=44096 (accessed 31 October 2013).
105. George H. W. Bush, "Address before a joint session of the Congress on the Persian Gulf crisis and the federal budget deficit", 11 September 1990, in ibid. Available at www.presidency.ucsb.edu/ws/?pid=18820 (accessed 31 October 2013).
106. William J. Clinton, "Statement on petroleum imports and energy security", 16 February 1995, in ibid. Available at www.presidency.ucsb.edu/ws/?pid=50988 (accessed 31 October 2013).
107. Yale Center for Environmental Law and Policy, "Yale environmental poll", Environmental Attitudes and Behavior Project, Yale Center for Environmental Law and Policy (2007). Available at http://envirocenter.yale.edu/uploads/epoll/YaleEnvironmentalPoll2007Keyfindings.pdf (accessed 16 July 2014).
108. Nik Nanos, "Entering the energy & environment policy frontier: an examination of the intersection of public opinion and public policy", Woodrow Wilson International Center for Scholars, 26 August 2013, pp. 17, 25. Available at www.

wilsoncenter.org/article/new-report-entering-the-energy-environment-policy-frontier (accessed 16 July 2014).

109. See, for example, the list of quotes from the 2008 presidential campaign compiled in Bryce, *Gusher of Lies*, pp. 4–7.
110. Daniel Yergin, "Ensuring energy security", *Foreign Affairs* 85/2 (2006), p. 76. Also see pp. 105–8 below.
111. Charles Krauthammer, "Energy independence? A serious plan requires taxes, ANWR and nukes", *Washington Post*, 26 January 2007.
112. John Hofmeister, *Why We Hate the Oil Companies: Straight Talk from an Energy Insider* (New York, 2010), pp. 25, 27.
113. Paul Stevens, "The 'shale gas revolution': developments and changes", Briefing Paper EERG BP 2012/04, Chatham House (2012), p. 7. Available at www.chathamhouse.org/publications/papers/view/185311 (accessed 9 December 2013).
114. See the different scenarios in US Energy Information Administration, *Annual Energy Outlook 2014*, pp. ES-2, ES-3.
115. Scott D. Sheffield, chief executive of Pioneer Natural Resources. Quoted in Clifford Krauss and Eric Lipton, "US inches toward goal of energy independence", *New York Times*, 22 March 2012.
116. Tim Mullaney, "Energy independence isn't just a pipe dream", *USA Today*, 16 May 2012.
117. See, for example, Michael Levi, "Think again: the American energy boom", *Foreign Policy* (July/August 2012), p. 56.
118. Edward L. Morse, *Energy 2020: Independence Day*, Citi GPS: Global Perspectives & Solutions (2013). Available at https://www.citivelocity.com/citigps/ReportSeries.action?recordId=16 (accessed 9 December 2013).
119. George W. Bush, "The President's news conference with European leaders in Vienna, Austria", 21 June 2006, in Peters and Woolley, *The American Presidency Project*. Available at www.presidency.ucsb.edu/ws/?pid=177 (accessed 9 December 2013).
120. George W. Bush, "The President's news conference", 21 March 2006, in ibid. Available at www.presidency.ucsb.edu/ws/?pid=65409 (accessed 9 December 2013).
121. For instance, in Barack H. Obama, "The President's weekly address", 7 February 2009, in ibid. Available at www.presidency.ucsb.edu/ws/?pid=85738 (accessed 9 December 2013).
122. The most important measure to reduce US oil consumption was the Corporate Average Fuel Economy (CAFE) regulations, first enacted by Congress in 1975.
123. Vietor, *Energy Policy in America*, p. 313.
124. US House Committee on Ways and Means, *Hearings on the Energy Crisis and Proposed Solutions*, 94th Cong., 1st sess., 3–17 March 1975, p. 27 (statement of William E. Simon, secretary of the Treasury).

Chapter 3 The Dangers of Foreign Oil

1. See the discussion on pp. 162–5.
2. US House Committee on International Relations, Special Subcommittee on Investigations, *Oil Fields as Military Objectives: A Feasibility Study*, report prepared by the Congressional Research Service, 94th Cong., 1st sess., Committee Print 56 (1975), p. 5.
3. Barack H. Obama, "The President's weekly address", 4 July 2009, in Peters and Woolley, *The American Presidency Project*. Available at www.presidency.ucsb.edu/ws/?pid=86356 (accessed 16 December 2013).
4. These questions are regularly asked within the existing literature on US foreign oil dependence – for example, in works attempting to calculate the "true" costs of foreign oil dependence. See, for example, Duffield, *Over a Barrel*. Similarly, the authors of an extensive study on dependency by the RAND Corporation state that the purpose of their work is to "critically evaluate commonly suggested links between imported oil and U.S. national security and to assess the costs and benefits of potential policies for reducing U.S. consumption and imports of oil". Crane et al., *Imported Oil*, p. iii.
5. See Campbell, *Writing Security*, p. 2.
6. This is a substantive assumption of the critical constructivist approach, on which the argument in this book is based. It represents a departure from classical concepts of national security, which posit the existence of actors with fixed or given identities in an environment where they experience objective and external threats. An excellent summary of this position is to be found in Weldes et al., "Constructing insecurity", quote on p. 10.
7. Weldes, *Constructing National Interests*, p. 13.
8. Authors in this tradition argue that security can refer to various different referent objects besides the traditional focus on the military security of the State. Among these are constituting political principles such as sovereignty, ideological systems, collective identities or natural environments. See Barry Buzan, Ole Wæver and Jaap De Wilde, *Security: A New Framework for Analysis* (Boulder, CO, 1998), pp. 21–3.
9. US Cabinet Task Force on Oil Import Control, *The Oil Import Question*, p. 30.
10. From a foreign policy point of view, it is often argued that the necessity to ensure the security of the global energy infrastructure, most importantly in the Persian Gulf region, requires substantial military forces at great cost to the taxpayer. See Crane et al., *Imported Oil*, ch. 5. From the point of view of the US military, another key issue revolves around the military's dependence on oil for fuel, which creates important financial and logistical problems. See CNA, *Powering America's Defense: Energy and the Risks to National Security*, report by the CNA Military Advisory Board (2009). Available at www.cna.org/reports/energy (accessed 21 December 2013).
11. US Cabinet Task Force on Oil Import Control, *The Oil Import Question*, p. 30.
12. Klare, *Blood and Oil*, pp. xiii–xiv.

13. Rutledge, *Addicted to Oil*, pp. 78–9.
14. Beaubouef, *Strategic Petroleum Reserve*, p. 220.
15. Daniel T. Rodgers, "Exceptionalism", in Anthony Mohlo and Gordon S. Wood (eds), *Imagined Histories: American Historians Interpret the Past* (Princeton, NJ 1998), pp. 22–3.
16. Halper and Clarke, *The Silence of the Rational Center*, pp. 21–3.
17. This argument is based on the view that identity is always established in relation to difference imagined as hostile or threatening. See pp. 116–17 below.
18. Campbell, *Writing Security*, p. 12.
19. US Cabinet Task Force on Oil Import Control, *The Oil Import Question* pp. 125–9.
20. On Arab warnings to the United States, see Andrew S. Cooper, *The Oil Kings How the US, Iran, and Saudi Arabia Changed the Balance of Power in the Middle East* (New York, 2011), pp. 110–17. On the two previous embargos, see Crane et al., *Imported Oil*, pp. 26–7.
21. "The oil weapon", *Wall Street Journal*, 19 October 1973.
22. "Calling the oil bluff", *Wall Street Journal*, 16 October 1973.
23. Nye, *Consuming Power*, p. 226.
24. Robert W. Tucker, "Oil: the issue of American intervention", *Commentary* January 1975, p. 30.
25. "Oil: Trying to cope with the looming crisis", *Time*, 14 October 1974.
26. "Faisal and oil", *Time*, 6 January 1975.
27. US Senate Committee on Banking, Housing and Urban Affairs, *Hearing on S. 2532 to Establish the Energy Independence Authority*, 94th Cong., 2nd sess., 12, 14 and 18 April and 10 May 1976, p. 439 (statement of James E. Akins, former US ambassador to Saudi Arabia).
28. McAlister, *Epic Encounters*, p. 236.
29. Linda Grant and Robert A. Rosenblatt, "Gas rationing scare: where do we stand?", *Los Angeles Times*, 8 December 1978.
30. See Thomas J. McCormick, *America's Half-Century: United States Foreign Policy in the Cold War and After* (Baltimore, MD, 1995), ch. 7.
31. Tucker, "The issue of intervention", p. 30.
32. Stephen D. Krasner, *Structural Conflict: The Third World against Global Liberalism* (Berkeley, CA, 1985), p. 71. Also see Walter J. Levy, "Oil and the Decline of the West", *Foreign Affairs* 58/5 (1980), pp. 999–1,015.
33. Gerald R. Ford, "Address to the nation on energy programs", 27 May 1975, in Peters and Woolley, *The American Presidency Project*. Available at www.presidency.ucsb.edu/ws/?pid=4942 (accessed 20 December 2013).
34. See Zaretsky, *No Direction Home*, p. 79.
35. Ford, "Special message to the Congress on energy", 7 January 1977.
36. Felix Ciută, "Conceptual notes on energy security: total or banal security", *Security Dialogue* 41/2 (2010), p. 132.
37. "The urgent need for new energy", *Los Angeles Times*, 24 October 1973.

38. US Senate Committee on Banking, Housing and Urban Affairs, *Energy Independence Authority*, pp. 3, 14 (statement of Nelson A. Rockefeller, vice-president of the United States).
39. "Gambling on oil", *Washington Post*, 21 February 1974.
40. Grant and Rosenblatt, "Gas rationing scare: where do we stand?"
41. Hobart Rowen, "OPEC's money grab", *Washington Post*, 19 April 1979.
42. Tucker, "The issue of intervention", pp. 23–4.
43. See Potter, *People of Plenty*, ch. 4.
44. Jack N. Anderson, "Oil puppets", *Washington Post*, 10 June 1979.
45. US Senate Committee on Banking, Housing and Urban Affairs, *Energy Independence Authority*, pp. 88–9 (statement of Walt W. Rostow, professor of economics and history at the University of Texas).
46. US House Committee on Ways and Means, *Hearings on the Energy Crisis*, p. 2 (statement of Albert C. Ullman, chairman of the Committee on Ways and Means).
47. Grant and Rosenblatt, "Gas rationing scare: where do we stand?"
48. Zaretsky, *No Direction Home*, pp. 83–7, quote on pp. 83–4.
49. Ford, "Address to the nation on energy and economic programs", 13 January 1975.
50. Ronald W. Reagan, "Address before a joint session of Congress on the State of the Union", 4 February 1986, in Peters and Woolley, *The American Presidency Project*. Available at www.presidency.ucsb.edu/ws/?pid=36646 (accessed 19 December 2013).
51. Alsop, "Oil blackmail threatens US independence."
52. Henry A. Kissinger, "The Washington Energy Conference: interdependence and cooperative solutions, 11 February 1974", *Vital Speeches of the Day* 40/11 (1974), p. 323.
53. Yergin, *The Prize*, p. 575.
54. US Senate Committee on Energy and Natural Resources, Subcommittee on Energy Regulation, *Hearing on S. 1134, S. 1205, S. 1417 and S. 1470 to Limit Oil Imports*, 96th Cong., 1st sess., 18 September 1979, p. 63 (statement of Thomas F. Eagleton, US senator from Missouri).
55. For a detailed discussion of the concept of energy interdependence see pp. 120–33.
56. Walter Laqueur and Edward Luttwak, "Oil", *Commentary* (October 1973), p. 40.
57. "Calling the oil bluff", *Wall Street Journal*, 16 October 1973.
58. Richard M. Nixon, "The President's news conference", 5 September 1973, in Peters and Woolley, *The American Presidency Project*. Available at www.presidency.ucsb.edu/ws/?pid=3948 (accessed 20 December 2013).
59. For example: "We have become increasingly at the mercy of others for the fuel on which our entire economy runs." Ford, "Address to the nation on energy programs", 27 May 1975.
60. Anderson, "Oil puppets."

61. "Pavlovian politics and Arab oil", *Washington Post*, 21 March 1974.
62. Cartoon by Buescher, *Chicago Defender*, 21 March 1974. Copyright KFS/Distr. Bulls.
63. Simon Head, "The monarchs of the Persian Gulf", *The New York Review of Books*, 21 March 1974.
64. Frank Vogl, "US misses its chance to influence OPEC", *Los Angeles Times*, 12 December 1978.
65. "The end of the embargo?", *Washington Post*, 15 March 1974.
66. James E. Carter, "National Energy Plan: address to the nation", 8 November 1977, in Peters and Woolley, *The American Presidency Project*. Available at www.presidency.ucsb.edu/ws/?pid=6904 (accessed 20 December 2013).
67. Zaretsky, *No Direction Home*, p. 79.
68. See p. 61 above.
69. Mobil Oil Company, "A fourth of July speech in search of an orator", advertisement published in the *New York Times*, 3 July 1975, in *ExxonMobil Historical Collection*, Dolph Briscoe Center for American History, University of Texas at Austin, part 2, box 2 207/E85.
70. Walter J. Hickel, "The energy war II: what we must do at home", *Reader's Digest*, February 1974, p. 102.
71. Keohane describes international hegemony as a "preponderance of material resources" and includes among these raw materials, sources of capital, control over markets and competitive advantages in the production of certain important goods. Oil enters this equation as the key raw material for industrialized economies. Robert O. Keohane, *After Hegemony: Cooperation and Discord in the World Political Economy* (Princeton, NJ, 2005), p. 32. Also see the discussion in Bromley, *American Hegemony*, ch. 1.
72. David S. Painter, "International oil and national security", *Daedalus* 120/4 (1991), p. 184.
73. Yergin, *The Prize*, p. 635.
74. Hans J. Morgenthau and W. Kenneth Thompson, *Politics Among Nations: The Struggle for Power and Peace* (New York, 1985), pp. 133–5.
75. ". . . or else", *Chicago Tribune*, 3 December 1973.
76. Thomas O. Enders, "OPEC and the industrial countries: the next ten years", *Foreign Affairs* 53/4 (1975), p. 625.
77. Klare, *Blood and Oil*, pp. 9–10.
78. Hobart Rowen, "A do-nothing oil policy", *Washington Post*, 22 February 1979.
79. Kissinger, "Interdependence and cooperative solutions", pp. 322–3.
80. Nixon, "Address to the nation about policies to deal with the energy shortages", 7 November 1973.
81. US Senate Committee on Banking, Housing and Urban Affairs, *Energy Independence Authority*, p. 3 (statement of Nelson A. Rockefeller, vice-president of the United States).
82. Hobart Rowen, "The 'economic agression' of the Arabs", *Washington Post*, 6 December 1973; "Arabs, Israel, and oil", *Los Angeles Times*, 6 September 1973.

83. Bush, "Address before a joint session of the Congress on the Persian Gulf crisis and the federal budget deficit", 11 September 1990.
84. See Yergin, *The Quest*, chs 6, 8.
85. See, for example, US Senate Committee on Energy and Natural Resources, *Hearing on the Geopolitics of Oil and Its Implications for US Economic and National Security*, 110th Cong., 1st sess., 10 January 2007, p. 25 (statement of Robert D. Hormats, vice-chairman of Goldman Sachs).
86. See US Senate Committee on Foreign Relations, Subcommittee on International Economic Policy, Export and Trade Promotion, *Hearing on Overview of Global Energy Security Issues*, 108th Cong., 1st sess., 8 April 2003, p. 3 (statement of Kyle McSlarrow, deputy secretary of energy).
87. For an example, see Daniel Yergin, "Thirty years of petro-politics", *Washington Post*, 17 October 2003.
88. Frederick W. Smith and Paul X. Kelley, "Are we ready for the next oil shock?", *Washington Post*, 11 August 2006.
89. Klare, *Blood and Oil*, pp. xiii–xiv. Emphasis in the original. Rutledge makes a similar argument in Rutledge, *Addicted to Oil*, pp. 78, 153.
90. Robert J. Samuelson, "Now do we get serious on oil?", *Washington Post*, 11 October 2001.
91. George W. Bush, "Address before a joint session of the Congress on the State of the Union", 23 January 2007, in Peters and Woolley, *The American Presidency Project*. Available at www.presidency.ucsb.edu/ws/?pid=24446 (accessed 1 January 2014).
92. US Senate Committee on Foreign Relations, *Hearing on Energy Security: Historical Perspectives and Modern Challenges*, 111th Cong., 1st sess., 12 May 2009, pp. 9–10 (statement of James E. Carter, former president of the United States).
93. George W. Bush, "Remarks on energy in Athens, Alabama", 21 June 2007, in Peters and Woolley, *The American Presidency Project*. Available at www.presidency.ucsb.edu/ws/?pid=75467 (accessed 1 January 2014).
94. George W. Bush, "Remarks on energy in Wilmington, Delaware", 24 January 2007, in ibid. Available at www.presidency.ucsb.edu/ws/?pid=24468 (accessed 1 January 2014).
95. Gerald F. Seib, "How oil dependence fuels US policies", *Wall Street Journal*, 22 August 2005.
96. Courtland Milloy, "Driven to war on a full tank of questions", *Washington Post*, 21 October 2001.
97. US Senate Committee on Foreign Relations, Subcommittee on Near Eastern and South Asian Affairs, *Hearing on US Foreign Policy, Petroleum, and the Middle East*, 109th Cong., 1st sess., 20 October 2005, p. 40 (statement of Tom Z. Collina, executive director of 20/20 Vision).
98. In 2009 dollars. See Crane et al., *Imported Oil*, pp. 63–5.
99. Klare, *Blood and Oil*, p. 7.

100. US Senate Committee on Foreign Relations, Subcommittee on Near Eastern and South Asian Affairs, *Hearing on US Foreign Policy, Petroleum, and the Middle East*, p. 26 (statement of Gal Luft, executive director of the Institute for the Analysis of Global Security).
101. Ibid., p. 48 (statement of Tom Z. Collina, executive director of 20/20 Vision).
102. Edward L. Morse, *Energy 2020: Independence Day*, p. 9.
103. US Cabinet Task Force on Oil Import Control, *The Oil Import Question*, p. 129.
104. US National Energy Policy Development Group, *National Energy Policy* (Washington, DC, 2001), ch. 8, p. 11.
105. Klare, *Blood and Oil*, p. 15.
106. Ford, "Address to the nation on energy programs", 27 May 1975.
107. US Energy Information Administration, "How dependent are we on foreign oil?", last updated 10 May 2013. Available at www.eia.gov/energy_in_brief/article/foreign_oil_dependence.cfm (accessed 16 July 2014).
108. Crane et al., *Imported Oil*, p. 59.
109. Smith and Kelley, "Are we ready for the next oil shock?"
110. Timothy E. Wirth, C. Boyden Gray and John D. Podesta, "The future of energy policy", *Foreign Affairs* 82/4 (2003), p. 134.
111. US Senate Committee on Energy and Natural Resources, Subcommittee on Energy Regulation, *Hearing on S. 1134, S. 1205, S. 1417 and S. 1470 to Limit Oil Imports*, p. 98 (statement of Morris A. Adelman, professor of economics at the Massachusetts Institute of Technology).
112. Yergin, "Ensuring energy security", p. 78.
113. Ciută, "Energy security", pp. 135–8, quote on p. 135. Emphasis in the original.
114. For an early example of these arguments, see Joseph S. Nye, "Energy nightmares", *Foreign Policy* (Autumn 1980), pp. 132–54.
115. US Senate Committee on Energy and Natural Resources, *Hearing on the Geopolitics of Oil*, p. 26 (statement of Robert D. Hormats, vice-chairman of Goldman Sachs).
116. Ann Bordetsky et al., *Securing America: Solving Our Oil Dependence Through Innovation*, National Resources Defense Council Issue Paper, 2005, p. 11. Available at www.nrdc.org/air/transportation/oilsecurity/plan.pdf (accessed 1 January 2014).
117. Wirth, Gray and Podesta, "The future of energy policy", p. 134.
118. Bush, "Remarks on energy in Wilmington, Delaware", 24 January 2007.
119. Conrad R. Burns, "Don't buy rogue oil", *Wall Street Journal*, 8 April 2002.
120. US Senate Committee on Foreign Relations, Subcommittee on Near Eastern and South Asian Affairs, *Hearing on US Foreign Policy, Petroleum, and the Middle East*, pp. 26–9 (statement of Gal Luft, executive director of the Institute for the Analysis of Global Security).
121. George W. Bush, "Remarks on energy", 18 June 2008, in Peters and Woolley, *The American Presidency Project*. Available at www.presidency.ucsb.edu/ws/?pid=77512 (accessed 15 July 2014).

22. George W. Bush, "Remarks during a meeting on energy security and climate change", 28 September 2007, in ibid. Available at www.presidency.ucsb.edu/ws/?pid=75839 (accessed 1 January 2014).
23. Bush, "Address before a joint session of the Congress on the State of the Union", 23 January 2007.
24. Epstein, *The Power of Words*, p. 11.
25. As Venn writes, "While the rapid and considerable increase in oil prices undoubtedly had an adverse effect upon the world economy, at the same time the rise in oil prices was a result, as well as a cause of the problems in the world economy, which was already showing signs of difficulty long before the 1973 oil crisis." Venn, *Oil Crisis*, ch. 5, quote on p. 149.
26. Crane et al., *Imported Oil*, p. 78.
27. See Bromley, *American Hegemony*, pp. 242–4; Kenneth N. Waltz, *Theory of International Politics* (New York, 1979), pp. 153–6.
28. See, for example, Crane et al., *Imported Oil*, ch. 4; Taylor and Van Doren, "The energy security obsession", pp. 480–1.

Chapter 4 The Foreignness of Foreign Oil

1. Klare, *Blood and Oil*, p. 18. Emphasis in the original.
2. US Cabinet Task Force on Oil Import Control, *The Oil Import Question*, p. 95.
3. US Federal Energy Administration, *Project Independence Report*, pp. 30, 361.
4. Melvin A. Conant, *The Oil Factor in US Foreign Policy, 1980–1990* (Lexington, KY, 1982), p. 111.
5. Ronald W. Reagan, "Radio address to the nation on oil prices", 19 April 1986, in Peters and Woolley, *The American Presidency Project*. Available at www.presidency.ucsb.edu/ws/?pid=37156 (accessed 3 January 2014).
6. The White House, *A National Security Strategy for a New Century* (1998), pp. 32–3. Available at www.au.af.mil/au/awc/awcgate/nss/nssr-1098.pdf (accessed 3 January 2014).
7. US National Energy Policy Development Group, *National Energy Policy*, ch. 8, pp. 8–9.
8. Data from tables 3.3a and 3.3d in US Energy Information Administration, *Monthly Energy Review: May 2014*, pp. 41, 45.
9. See Campbell, *Writing Security*, pp. 8–12.
10. Benedict Anderson, *Imagined Communities: Reflections on the Origin and Spread of Nationalism* (London, 1991).
11. Connolly, *Identity|Difference*, p. 64. Also see Weldes et al., "Constructing insecurity".
12. Iran has often been included in this group, despite the fact that it is not an Arab nation and had no role in the oil embargo of 1973.

13. Gearóid Ó Tuathail and John Agnew, "Geopolitics and discourse: practica geopolitical reasoning in American foreign policy", *Political Geography* 11/ (1992), p. 192.
14. Edward W. Said, *Orientalism* (London, 2003), p. 54.
15. Ibid., p. 300.
16. McAlister, *Epic Encounters*, pp. 4–5. For Shapiro's original formulation of th concept, see Michael J. Shapiro, "Moral geographies and the ethics of post sovereignty", *Public Culture* 6 (1994), pp. 479–502.
17. Said, *Orientalism*, pp. 285–6. Said also discusses these issues in detail in a late book: Edward W. Said, *Covering Islam: How the Media and the Experts Determin How We See the Rest of the World* (New York, 1997), ch. 2.
18. McAlister, *Epic Encounters*, pp. 126, 136.
19. Zaretsky, *No Direction Home*, p. 95.
20. A detailed account can be found in David A. Baldwin, "Interdependence an power: a conceptual analysis", *International Organization* 34/4 (1980) pp. 471–506.
21. Dale C. Copeland, "Economic interdependence and war: a theory of trad expectations", *International Security* 20/4 (1996), p. 12. The dispute is als connected to general disagreement about the importance of power in stat behavior, the idea of relative versus absolute gains and other issues in the so called neo–neo debate. See, for example, Robert O. Keohane and Josep S. Nye, *Power and Interdependence* (New York, 2001), ch. 1; Kenneth N. Waltz "The myth of national interdependence", in Charles P. Kindleberger (ed.), *Th International Corporation* (Cambridge, MA, 1970); Richard Rosecrance an Arthur Stein, "Interdependence: myth or reality?", *World Politics* 24/1 (1973) pp. 1–27.
22. Richard Rosecrance, *The Rise of the Trading State: Commerce and Conquest in th Modern World* (New York, 1986). For a summary of the liberal position, se Susan M. McMillan, "Interdependence and conflict", *Mershon Internationa Studies Review* 41/1 (1997), pp. 35–40; Copeland, "Economic interdependenc and war", pp. 8–9.
23. John J. Mearsheimer, "Disorder restored", in Graham Allison and Gregor F. Treverton (eds), *Rethinking America's Security: Beyond Cold War to New Worl Order* (New York, 1992), pp. 222–3.
24. McMillan, "Interdependence and conflict", pp. 40–2; Copeland, "Economi interdependence and war", pp. 10–11.
25. Waltz, *Theory of International Politics*, p. 153.
26. Rosecrance, *The Rise of the Trading State*, p. 13.
27. Joseph S. Nye, *Understanding International Conflicts: An Introduction to Theor and History* (New York, 2009), pp. 221–6; Keohane, *After Hegemony*, chs 9–10 i particular. Also see the concept of "complex interdependence" in Keohan and Nye, *Power and Interdependence*.
28. See the discussion on pp. 90–3.
29. Waltz, *Theory of International Politics*, p. 144.

30. United Nations General Assembly, "A/RES/S-6/3201, resolution adopted by the General Assembly 3201 (S-VI): declaration on the establishment of a new international economic order", 1 May 1974. Available at www.un-documents.net/s6r3201.htm (accessed 17 July 2014). Cf. Krasner, *Structural Conflict*, p. 7.
31. Parra, *Oil Politics*, p. 191.
32. Kissinger, "Interdependence and cooperative solutions", p. 325.
33. See McCormick, *America's Half-Century*, pp. 195–203.
34. Tucker, "The issue of intervention", p. 31.
35. Stanley Hoffmann, "Letter to the editor", *Commentary* (April 1975), p. 6.
36. Robert W. Tucker, "Letter to the editor", *Commentary* (April 1975), p. 21.
37. Yet, this liberal vision was coupled with the rather mercantilist idea that American companies should have direct ownership of foreign oil resources. Emily S. Rosenberg, *Spreading the American Dream: American Economic and Cultural Expansion, 1890–1945* (New York, 1982), pp. 125–7.
38. See Vietor, *Energy Policy in America*, ch. 5.
39. US Cabinet Task Force on Oil Import Control, *The Oil Import Question*, pp. 4–6.
40. US Senate Committee on Interior and Insular Affairs, *Hearing on the President's Energy Message and S. 1570*, 93rd Cong., 1st sess., 1 May 1973, p. 140 (statement of William J. Casey, under secretary of state for economic affairs).
41. McAlister, *Epic Encounters*, p. 137.
42. US Senate Committee on Foreign Relations, *Hearing on the Implications of the Current Energy Problem for United States Foreign Policy*, 93rd Cong., 1st sess., 30 and 31 May 1973, p. 26 (statement of David Freeman, Ford Foundation Energy Policy Project). Also see the response by J. William Fulbright, chairman of the Committee on Foreign Relations, in ibid., p. 36.
43. "President Ford on Energy", *Wall Street Journal*, 17 January 1975.
44. US Senate Committee on Banking, Housing and Urban Affairs, *Energy Independence Authority*, p. 3 (statement of Nelson A. Rockefeller, vice-president of the United States).
45. Clayton Fritchey, "'Project Independence': freedom from OPEC", *Washington Post*, 11 June 1979.
46. Hodel, "National Energy Policy Plan", p. 133.
47. Pietro S. Nivola, "Energy independence or interdependence? Integrating the North American energy market", *The Brookings Review* 20/2 (2002), pp. 24, 27.
48. US National Energy Policy Development Group, *National Energy Policy*, ch. 8, pp. 4–8.
49. George W. Bush, "The President's news conference with Summit of the Americas leaders in Quebec City", 22 April 2001, in Peters and Woolley, *The American Presidency Project*. Available at www.presidency.ucsb.edu/ws/?pid=45638 (accessed 4 January 2014).
50. US Cabinet Task Force on Oil Import Control, *The Oil Import Question*, pp. 6, 35.
51. US National Energy Policy Development Group, *National Energy Policy*, ch. 8, p. 4.
52. See, for example, Morse, *Energy 2020: Independence Day*.

53. "Oil and unemployment", *Washington Post*, 18 November 1973.
54. Samuelson, "Now do we get serious on oil?"
55. Copeland, "Economic interdependence and war", p. 12.
56. See, for example, Taylor and Van Doren, "The energy security obsession", pp. 477–9; Philip E. Auerswald, "The irrelevance of the Middle East", *The American Interest* 2/5 (2007), pp. 21–2; Morris A. Adelman, *The Genie out of the Bottle: World Oil since 1970* (Cambridge, MA, 1995).
57. Giacomo Luciani, "Oil and political economy in the international relations of the Middle East", in Louise Fawcett (ed.), *International Relations of the Middle East* (Oxford, 2009), p. 90.
58. See Eric M. Uslaner, "Energy policy and free trade in Canada", *Energy Policy* 17/4 (1989), pp. 323–30.
59. On this policy see Rutledge, *Addicted to Oil*, pp. 80–5; Randall, *Foreign Oil Policy*, pp. 292–3.
60. Donald Alper, "Oil: the rationale for Canada's hard line", *Los Angeles Times*, 7 January 1975.
61. Nye, *Understanding International Conflicts*, p. 209.
62. This theoretical framework is laid out in Alexander Wendt, "Anarchy is what states make of it: the social construction of power politics", *International Organization* 46/2 (1992), pp. 391–425.
63. R. Emmett Tyrrell, "Chimera in the Middle East", *Harper's* (November 1976), p. 38.
64. US Senate Committee on Foreign Relations, Subcommittee on Near Eastern and South Asian Affairs, *Hearings on Middle East Peace Prospects*, 94th Cong., 2nd sess., 19 May; 7, 21, 30 June; 19 and 26 July 1976, pp. 207, 222 (statement of Edward Luttwak, associate director of the Washington Center for Foreign Policy Research).
65. There is plenty of literature on this issue. See, for example, Matthew F. Jacobs, *Imagining the Middle East: The Building of an American Foreign Policy, 1918–1967* (Chapel Hill, NC, 2011); Douglas Little, *American Orientialism: The United States and the Middle East since 1945* (Chapel Hill, NC, 2008), ch. 1; Michael B. Oren, *Power, Faith and Fantasy: America in the Middle East, 1776 to the Present* (New York, 2007).
66. McAlister, *Epic Encounters*; Catherine A. Lutz and Jane L. Collins, *Reading National Geographic* (Chicago, 1993).
67. Jacobs, *Imagining the Middle East*; Said, *Covering Islam*; Yahya R. Kamalipour (ed.), *The US Media and the Middle East: Image and Perception* (Westport, CT, 1995).
68. Said, *Orientalism*, pp. 284–5, 300.
69. Cf. McAlister, *Epic Encounters*, p. 126.
70. Edward W. Said, *Culture and Imperialism* (New York, 1994), p. 261.
71. A good account of this history appears in Little, *American Orientialism*, ch. 1.
72. Said, *Orientalism*, p. 287.
73. Jack G. Shaheen, *Reel Bad Arabs: How Hollywood Vilifies a People* (Northampton, MA, 2001), p. 2.

4. Jack G. Shaheen, *Guilty: Hollywood's Verdict on Arabs after 9/11* (Northampton, MA, 2008).
5. See Little, *American Orientialism*, ch. 1.
6. Anne McClintock, *Imperial Leather: Race, Gender and Sexuality in the Colonial Contest* (New York, 1995), p. 40.
7. CIA Report SR-13 "Arab world", 27 September 1949. Quoted in Little, *American Orientialism*, p. 26.
8. US Department of State, "The roots of Arab resistance to modernization", 12 September 1969. Quoted in Jacobs, *Imagining the Middle East*, pp. 140–1.
9. On this point see Little, *American Orientialism*, pp. 25–8.
0. US Department of State, *Foreign Relations of the United States (FRUS): 1958–1960, Dwight D. Eisenhower, Volume XII (Near East Region, Iraq, Iran, Arabian Peninsula)* (Washington, DC, 1993), p. 99.
1. See Parra, *Oil Politics*, pp. 6–10; Paul Stevens, "National oil companies and international oil companies in the Middle East: under the shadow of government and the resource nationalism cycle", *Journal of World Energy Law & Business* 1/1 (2008), pp. 9–11.
2. Bromley, *American Hegemony*, p. 118.
3. Painter, *Oil and the American Century*, p. 206.
4. Ibid., p. 153.
5. Quoted in Rosenberg, *Spreading the American Dream*, p. 128. Emphasis added.
6. See, for instance, US Department of State, *Foreign Relations of the United States (FRUS): 1945–1953, Harry S Truman, Volume VIII (The Near East and Africa)* (Washington, DC, 1969), p. 916.
7. US Senate Committee on Energy and Natural Resources, Subcommittee on Energy Regulation, *Hearing on S. 1134, S. 1205, S. 1417 and S. 1470 to Limit Oil Imports*, p. 59 (statement of John A. Durkin, US senator from New Hampshire).
8. Morris A. Adelman, "Is the oil shortage real? Oil companies as OPEC tax-collectors", *Foreign Policy* (Winter 1972/73), p. 71.
9. According to Yergin, the companies were anxious not to break ties with either the producing or consuming countries and thus decided to operate a system of "equal misery" in which they allocated the same percentage of cutbacks from total supply to all consuming countries. See Yergin, *The Prize*, pp. 601–6.
0. See Jacobs, *Imagining the Middle East*, pp. 142–51.
1. Rosenberg, *Spreading the American Dream*, p. 7. Also see Anders Stephanson, *Manifest Destiny: American Expansionism and the Empire of Right* (New York, 1996), p. 5.
2. Vitalis, *America's Kingdom*, pp. 9–10.
3. Charles L. Harding, "Middle East oil: its development and its impact on our national interest", speech delivered before the Harvard Engineering Society, 9 March 1950, in *ExxonMobil Historical Collection*, Dolph Briscoe Center for American History, University of Texas at Austin, part 2, box 2.207/E183, p. 1.

94. Chevron, *Standard Oil of California Bulletin*, 1946. Quoted in Vitalis, *America'* *Kingdom*, p. 35.
95. Rutledge, *Addicted to Oil*, p. 28; Vitalis, *America's Kingdom*, p. xiii. For a fictiona account of this encounter from the Arab perspective, see Abdelrahman Munif *Cities of Salt* (New York, 1987).
96. US Department of State, "The US and the impending energy crisis", report, 9 March 1972, Declassified Documents Reference System (CK3100516177).
97. Nye, *Understanding International Conflicts*, p. 221.
98. In essence, the term "obsolescing bargain" refers to the process by whicl multinational companies suffer losses when pushed to renegotiate contract with host governments after a majority of their investment has sunk and risl declined. Put most simply, over time the majors had to offer better terms to Middle Eastern governments. See George Joffé et al., "Expropriation of oil anc gas investments: historical, legal and economic perspectives in a new age o resource nationalism", *Journal of World Energy Law & Business* 2/1 (2009), p. 3 For the original formulation of Obsolescing Bargain Theory, see Raymonc Vernon, *Sovereignty at Bay: The Multinational Spread of U.S. Enterprise* (New York, 1971).
99. Fiona Venn, *Oil Diplomacy in the Twentieth Century* (New York, 1986), p. 131
100. Ibid., pp. 129, 138.
101. See Parra, *Oil Politics*, pp. 150–4. Peru (1968) and Bolivia (1969) hac nationalized oil-related assets even earlier.
102. David S. Painter, "Oil", in Alexander DeConde, Fredrik Logevall and Richarc Dean Burns (eds), *Encyclopedia of American Foreign Policy*, vol. 3 (New York 2002), p. 15.
103. J. E. Hartshorn, *Oil Trade: Politics and Prospects* (Cambridge, 1993), p. 140.
104. Venn, *Oil Diplomacy*, p. 136.
105. See, for example, "The Arabs' new oil squeeze: dimouts, slowdowns, chills", *Time*, 19 November 1973.
106. Tyrrell, "Chimera", p. 38.
107. Norman Kempster and Ronald J. Ostrow, "Oil: the tie that binds US, Arabs", *Los Angeles Times*, 22 November 1978.
108. McAlister, *Epic Encounters*, p. 139.
109. Levy, "Oil and the decline of the West", p. 1,014.
110. See, for example, William Raspberry, "Blackmail—or diplomacy?", *Washington Post*, 23 November 1973.
111. McAlister, *Epic Encounters*, p. 139.
112. Tibor R. Machan, "Letter to the editor", *Commentary* (April 1975), pp. 12–13.
113. William F. Buckley, "The case for the use of force to get Arab oil", *Los Angeles Times*, 10 October 1973.
114. Said, *Covering Islam*, p. 38.
115. Krasner, *Structural Conflict*, pp. 1–13.
116. Robert W. Tucker, "Further reflections on oil & force", *Commentary* (March 1975), p. 55.

17. Said, *Covering Islam*, p. 38.
18. G. John Ikenberry, *After Victory: Institutions, Strategic Restraint, and the Rebuilding of Order after Major Wars* (Princeton, NJ, 2001), p. 163.
19. Doug Stokes and Sam Raphael, *Global Energy Security and American Hegemony* (Baltimore, MD, 2010), p. 15.
20. Ibid., p. 19.
21. This view is taken by Bromley, *American Hegemony*, pp. 122–3. The same argument appears in critical accounts of the "American empire". See, for example, Noam Chomsky and Gilbert Achcar, *Perilous Power: The Middle East and US Foreign Policy: Dialogues on Terror, Democracy, War, and Justice* (Boulder, CO, 2007), p. 54.
22. Memo to President Harry S. Truman, prepared by the Chief of the Division of Near Eastern Affairs, August 1945, reprinted in US Department of State, *FRUS, Vol. VIII (1945–1953)*, p. 45.
23. Robert Latham, *The Liberal Moment: Modernity, Security, and the Making of Postwar International Order* (New York, 1997), p. 40.
24. On the Great Oil Deals, see Painter, *Oil and the American Century*, pp. 102–10.
25. See David S. Painter, "The Marshall Plan and oil", *Cold War History* 9/2 (2009), pp. 159–75.
26. Bromley, *American Hegemony*, p. 118.
27. Latham, *Liberal Moment*, pp. 14–17.
28. Ikenberry, *After Victory*, pp. 164–5.
29. Tucker, "The issue of intervention", p. 25.
30. Nye, "Energy narratives", pp. 77–9.
31. US Senate Committee on Banking, Housing and Urban Affairs, *Energy Independence Authority*, p. 7 (statement of Nelson A. Rockefeller, vice-president of the United States).
32. On American views about oil cartels, see Olien and Davids Olien, *Oil and Ideology*, chs 1–2. On OPEC, see Craig S. Karpel, "Ten ways to break OPEC", *Harper's* (January 1976).
33. See, for example, Raspberry, "Blackmail—or diplomacy?"
34. Gerald R. Ford, "Remarks to the Ninth World Energy Conference, Detroit, Michigan", 23 September 1974, in Peters and Woolley, *The American Presidency Project*. Available at www.presidency.ucsb.edu/ws/?pid=4732 (accessed 15 July 2014).
35. Stevens, "National oil companies", p. 5.
36. Jahangir Amuzegar, *Managing the Oil Wealth: OPEC's Windfalls and Pitfalls* (London, 2001), p. 204.
37. See Stevens, "National oil companies", pp. 17–23.
38. Ibid., p. 24.
39. See the argument in Joffé et al., "Expropriation of oil and gas investments", pp. 3–23.
40. Yergin, *The Prize*, p. 770.

141. American opposition to resource nationalism goes back to the very beginning of the hydrocarbon society. For accounts of early cases in Latin America an elsewhere, see Venn, *Oil Diplomacy*, ch. 4; Painter, *Oil and the American Centur* ch. 6.
142. US Energy Information Administration, *International Energy Outlook 200*. DOE/EIA-0484 (2002), pp. 26, 31. Available at www.eia.gov/forecast archive/ieo02/pdf/0484%282002%29.pdf (accessed 6 January 2014).
143. Rutledge, *Addicted to Oil*, p. 141.
144. See Stevens, "National oil companies", pp. 24–5.
145. Edward L. Morse and Amy Myers Jaffe, *Strategic Energy Policy Challenges for th 21st Century: Report of An Independent Task Force*, cosponsored by the Jame A. Baker III Institute for Public Policy of Rice University and the Council o Foreign Relations (2001), p. 22. Available at www.cfr.org/energy-policy/strategic energy-policy-challenges-21st-century/p3942 (accessed 6 January 2014).
146. US Senate Committee on Energy and Natural Resources, *Hearing on th Geopolitics of Oil*, pp. 36–7 (statement of Charles F. Wald, former deput commander of the United States European Command).
147. US National Energy Policy Development Group, *National Energy Policy*, ch. 8 p. 18.
148. See, for example, Stokes and Raphael, *Global Energy Security*, pp. 96–100.
149. Robert Stobaugh and Daniel Yergin, "After the second shock: pragmati energy strategies", *Foreign Affairs* 57/4 (1979), p. 839.
150. Seymour M. Lipset, "Some social requisites of democracy: economi development and political legitimacy", *American Political Science Review* 53/ (1959), pp. 69–105.
151. See Richard M. Auty, *Sustaining Development in Mineral Economies: The Resour Curse Thesis* (London, 1993).
152. For an overview of these troublesome relationships see, for example, Paul Steven and Evelyn Dietsche, "Resource curse: an analysis of causes, experiences an possible ways forward", *Energy Policy* 36/1 (2008), pp. 56–65; Andrew Rosse "The political economy of the resource curse: a literature survey", Workin Paper No. 268, Institute of Development Studies (IDS), Centre for the Futur State (2006). Available at www.ids.ac.uk/files/WP268.pdf (accessed 16 Jul 2014); Michael L. Ross, "The political economy of the resource curse", *Worl Politics* 51/2 (1999), pp. 297–322.
153. William L. Cleveland, *A History of the Modern Middle East* (Boulder, CO 2004), p. 441. Also see John L. Esposito, *The Islamic Threat: Myth or Reality* (New York, 1999), ch. 1.
154. George Joffé, "Confrontational mutual perceptions and images: orientalism and occidentalism in Europe and the Islamic world", *The International Spectator Italian Journal of International Affairs* 42/2 (2007), p. 167.
155. McAlister, *Epic Encounters*, p. 199.
156. On this point see Esposito, *The Islamic Threat*, pp. 1–4; Said, *Covering Islam* pp. 36–45.

157. Bernard Lewis, "The roots of Muslim rage", *Atlantic Monthly* (September 1990), p. 48.
158. The clash-of-civilizations thesis was originally formulated by Lewis in the above essay. It was later popularized by Samuel Huntington in Samuel P. Huntington, "The clash of civilizations?", *Foreign Affairs* 72/3 (1993), pp. 22–49.
159. George W. Bush, "Remarks on energy independence", 6 February 2003, in Peters and Woolley, *The American Presidency Project*. Available at www.presidency.ucsb.edu/ws/?pid=63784 (accessed 6 January 2014).
160. Richard B. Cheney, "Defending liberty in a global economy", speech at The Collateral Damage Conference, The Cato Institute, 23 June 1998. Available at www.cato.org/publications/speeches/defending-liberty-global-economy (accessed 6 January 2014).
161. Taylor and Van Doren, "The energy security obsession", pp. 478–80.
162. On the US–Saudi relationship over oil, see, for example, Klare, *Blood and Oil*, ch. 2.
163. Irwin M. Stelzer, "Can we do without Saudi oil?", *Weekly Standard*, 19 November 2001. Available at www.weeklystandard.com/Content/Public/Articles/000/000/000/525pggsa.asp (accessed 5 August 2014).
164. "Lethal embrace" is Klare's term. Klare, *Blood and Oil*, p. 26. The remaining quotes are from Doug Bandow, "Befriending Saudi princes: a high price for a dubious alliance", Policy Analyis No. 428, The Cato Institute, 2002, pp. 2, 5. Available at www.cato.org/pubs/pas/pa-428es.html (accessed 6 January 2014).
165. Quoted in Kempster and Ostrow, "Oil: the tie that binds".
166. These principles include the prohibition of interest (Arabic: *Riba*) and waste (*Israf*). See Øystein Noreng, "Islam and oil", in Eni S.P.A (Ente Nazionale Idrocarburi) (ed.), *Encyclopedia of Hydrocarbons (vol. IV: Economics, Policies, and Legislation)* (Rome, 2006), quote on p. 431. Also see Rutledge, *Addicted to Oil*, ch. 11; Øystein Noreng, *Oil and Islam: Social and Economic Issues* (Chichester, 1997).
167. Cleveland, *A History of the Modern Middle East*, p. 462. Also see the account in Ahmed Rashid, *Taliban: Militant Islam, Oil and Fundamentalism in Central Asia* (London, 2010).
168. Stelzer, "Can we do without Saudi oil?"
169. US Senate Committee on Energy and Natural Resources, *Hearing on Energy Independence*, 109th Cong., 2nd sess., 7 March 2006, p. 18 (statement of R. James Woolsey, former director of the Central Intelligence Agency).
170. See Michael H. Hunt, *Ideology and US Foreign Policy* (New Haven, CT, 2009), chs 2–3.
171. On CERT, see Winona LaDuke, *The Winona LaDuke Reader: A Collection of Essential Writings* (Stillwater, MN, 2002), pp. 116–29.
172. Quote in Molly Ivins, "120 tribes form new group to block Anti-Indian moves", *New York Times*, 16 April 1978. Also see George E. Simpson and J. Milton Yinger, *Racial and Cultural Minorities: An Analysis of Prejudice and Discrimination* (New York, 1985), pp. 425–7.

Chapter 5 Fighting for Oil

1. See Jacob Weisberg, "The corn is high", *New Republic* (12 November 1990), p. 12.
2. "Gulf protest breaks out in US Senate, gallery cleared", *Reuters News*, 11 January 1991.
3. Ghosh, "Petrofiction", p. 30.
4. For accounts of these struggles, see Mitchell, *Carbon Democracy*; Vitalis, *America's Kingdom*.
5. "Striking seaman invade Times Square", *New York Times*, 28 April 1939.
6. Note of Prime Minister Georges Clemenceau to President Woodrow Wilson, 15 December 1917, translation by the author. See Eric D. K. Melby, *Oil and the International System: The Case of France, 1918–1969* (New York, 1981), p. 10.
7. "Blood and oil title of next world tragedy", *Chicago Tribune*, 23 September 1923.
8. That oil and power are inherently connected is, for instance, the main thesis in Yergin, *The Prize*.
9. See, for example, Paul Collier and Anke Hoeffler, "Greed and grievance in civil war", *Oxford Economic Papers* 56/4 (2004), pp. 563–95; Michael T. Klare, *Resource Wars: The New Landscape of Global Conflict* (New York, 2001). A good overview of the literature can be found in Michael T. Klare, "There will be blood: political violence, regional warfare, and the risk of great-power conflict over contested energy sources", in Gal Luft and Anne Korin (eds), *Energy Security Challenges for the 21st Century: A Reference Handbook* (Santa Barbara, CA, 2009).
10. For example, in US National Security Council, "Long-range US policy toward the Near-East", draft statement of policy, NSC 5801, 10 January 1958, Declassified Documents Reference System (CK3100451753), p. 13.
11. Congressional Record, 1st sess., vol. 119, pt. 13 (21 May 1973), p. 16,263 (statement of Senator J. William Fulbright).
12. Cooper, *The Oil Kings*, pp. 102–5, 115.
13. Nixon, "The President's news conference", 5 September 1973. The government of Mohammad Mossaddegh was overthrown in 1953 in a coup orchestrated and financed in part by the United States. Iran's oil resources came under the control of an international consortium consisting of European and American oil companies. See Yergin, *The Prize*, ch. 23.
14. "After the oil embargoes", *Washington Post*, 4 December 1973.
15. Cooper, *The Oil Kings*, p. 130.
16. Secretary of Defense James Schlesinger on *Washington straight talk*, 7 January 1974. Quoted in US House Committee on International Relations, Special Subcommittee on Investigations, *Oil Fields as Military Objectives*, p. 80.
17. Marwan R. Buheiry, *US Threats of Intervention against Arab Oil, 1973–1979* (Beirut, 1980), p. 11.
18. Cooper, *The Oil Kings*, p. 132.

9. Ford, "Remarks to the Ninth World Energy Conference, Detroit, Michigan", 23 September 1974.
0. Buheiry, *US Threats of Intervention*, p. 24.
1. See Isidor F. Stone, "War for oil?", *New York Review of Books* (6 February 1975).
2. US Department of State, "Secretary Henry Kissinger interviewed for Business Week magazine", *Department of State Bulletin* 72/1857 (1975), p. 101.
3. Secretary of Defense Schlesinger's press conference at the Pentagon, 14 January 1975. Quoted in US House Committee on International Relations, Special Subcommittee on Investigations, *Oil Fields as Military Objectives*, p. 81.
4. "Gerald Ford: They will see something is being done", *Time*, 20 January 1975.
5. Gerald R. Ford, "The President's news conference", 21 January 1975, in Peters and Woolley, *The American Presidency Project*. Available at www.presidency.ucsb.edu/ws/?pid=5027 (accessed 1 January 2014).
6. Reported in Cooper, *The Oil Kings*, p. 129.
7. Buheiry, *US Threats of Intervention*, p. 26; Stone, "War for oil?"
8. Reported in Cooper, *The Oil Kings*, p. 128.
9. "Mr. Kissinger on oil", *Washington Post*, 6 January 1975.
0. Quoted in Clifton Daniel, "Kissinger remark on force sparks wide speculation", *New York Times*, 7 January 1975.
1. Stephen S. Rosenfeld, "Oil and military force", *Washington Post*, 10 January 1975; Tom Wicker, "Stating the obvious", *New York Times*, 12 January 1975; "Oil and force", *Wall Street Journal*, 10 January 1975.
2. Ernest Conine, "The Kissinger 'oil-war' flap: why all the excitement?", *Los Angeles Times*, 10 January 1975.
3. Cooper, *The Oil Kings*, p. 233.
4. Stefan Halper and Jonathan Clarke, *America Alone: The Neo-Conservatives and the Global Order* (Cambridge, 2004), p. 11. For a discussion of first-generation neo-conservatives, see ch. 2.
5. Buckley, "The use of force".
6. William F. Buckley, "And what has Europe done for America lately?", *Los Angeles Times*, 15 November 1973.
7. Tucker, "The issue of intervention"; Tucker, "Oil & force".
8. Tucker, "The issue of intervention", p. 28.
9. Miles Ignotus, "Seizing Arab oil", *Harper's* (March 1975).
0. Ibid., p. 45.
1. See Andrew Higgins, "Power and peril: America's supremacy and its limits", *Wall Street Journal*, 4 February 2004.
2. James E. Akins. Quoted in Robert Dreyfuss, "The thirty-year itch", *Mother Jones* (March/April 2003), p. 42.
3. Ignotus, "Seizing Arab oil", pp. 45–56 ("appeasement"), 57 ("OPEC members"), 58 ("near-impunity"), 60 ("Iranians"), 62 ("sigh of relief"), 60 ("let us have the oil").
4. Ibid., p. 62.

45. 1975 Survey conducted by *Time*. Quoted in Thomas C. Barger, "Suppose we seized Arab oil fields", *Los Angeles Times*, 13 April 1975.
46. Cooper, *The Oil Kings*, p. 92.
47. See the data in Bruce Russett and Miroslav Nincic, "American opinion on the use of military force abroad", *Political Science Quarterly* 91/3 (1976), pp. 411–31
48. Charles L. Schultze, "Oil war: fantasy no one needs", *Washington Post*, 26 January 1975.
49. See, for instance, James Chace, "American jingoism", *Harper's* (May 1976).
50. US House Committee on International Relations, Special Subcommittee on Investigations, *Oil Fields as Military Objectives*, p. 75.
51. Michael A. Palmer, *Guardians of the Gulf: A History of America's Expanding Role in the Persian Gulf, 1833–1992* (New York, 1992), pp. 100–1. See also "Military men challenge Mideast force strategy", *New York Times*, 10 January 1975.
52. Barger, "Suppose we seized Arab oil fields".
53. Cartoon created by Wayne Stayskal, *Chicago Tribune*, 11 February 1975. Copyright Tribune Content Agency, LLC. All Rights Reserved. Reprinted with permission.
54. This document was declassified following a Freedom of Information request by the author. James E. Akins, "War for oil", cable from the US Embassy in Jiddah to the State Department, 13 April 1975, National Archives and Records Administration, P-Reel Index (1975JIDDA A-23).
55. Ibid., quote on p. 15.
56. Congressional Record, 1st sess., vol. 121, pt. 1 (23 January 1975), p. 1,106 (statement of Senator Clifford P. Hansen).
57. "Letters to the editor", *New York Times*, 12 January 1975. Emphasis in the original.
58. "Letters to the editor", *Los Angeles Times*, 24 April 1975.
59. Congressional Record, 1st sess., vol. 119, pt. 13 (21 May 1973), p. 16,263 (statement of Senator William J. Fulbright); Buckley, "And what has Europe done?"
60. Tucker, "The issue of intervention", p. 24.
61. Ibid., p. 25.
62. A point that Luttwak had already made in an earlier article. See Laqueur and Luttwak, "Oil".
63. US Senate Committee on Foreign Relations, Subcommittee on Near Eastern and South Asian Affairs, *Hearings on Middle East Peace Prospects*, quotes on pp. 205–7 (statement of Edward N. Luttwak, associate director of the Washington Center for Foreign Policy Research).
64. Ignotus, "Seizing Arab oil", p. 48.
65. Congressional Record, 1st sess., vol. 121, pt. 1 (23 January 1975), p. 1,106 (statement of Senator Clifford P. Hansen).
66. Ibid., p. 1,103 (statement of Senator Jesse A. Helms).
67. "Letters to the Editor", *Washington Post*, 14 January 1975.

8. "Letters to the Editor", *New York Times*, 10 January 1975.
9. Congressional Record, 1st sess., vol. 121, pt. 1 (23 January 1975), p. 1,106 (statement of Senator James A. McClure).
0. Congressional Record, 1st sess., vol. 121, pt. 1 (21 January 1975), quotes on pp. 844–5 (statement of Senator Mark O. Hatfield).
1. "Letters to the Editor", *Harper's*, May 1975, p. 16.
2. Susan A. Brewer, *Why America Fights: Patriotism and War Propaganda from the Philippines to Iraq* (Oxford, 2009), pp. 3–4.
3. Stokes and Raphael, *Global Energy Security*, p. 220.
4. Congressional Record, 1st sess., vol. 121, pt. 1 (23 January 1975), p. 1,103 (statement of Senator Jesse A. Helms).
5. George H. W. Bush, "The President's news conference on the Persian Gulf crisis", 8 November 1990, in Peters and Woolley, *The American Presidency Project*. Available at www.presidency.ucsb.edu/ws/?pid=19019 (accessed 5 January 2014).
6. For a detailed account, see John R. Greene, *The Presidency of George Bush* (Lawrence, KS, 2000), chs 8–9; Palmer, *Guardians of the Gulf*, chs 8–12.
7. Brewer, *Why America Fights*, p. 4.
8. George H. W. Bush, "Remarks at the Annual Convention of the National Religious Broadcasters", 28 January 1991, in Peters and Woolley, *The American Presidency Project*. Available at www.presidency.ucsb.edu/ws/?pid=19250 (accessed 5 January 2014). On the administration's narrative, see Rachel Martin Harlow, "Agency and agent in George Bush's Gulf War rhetoric", in Martin J. Medhurst (ed.), *The Rhetorical Presidency of George H. W. Bush* (College Station, TX, 2006); Philip M. Taylor, *War and the Media: Propaganda and Persuasion in the Gulf War* (Manchester, 1998); David Campbell, *Politics Without Principle: Sovereignty, Ethics, and the Narratives of the Gulf War* (Boulder, CO, 1993).
9. These different voices are documented in Micah L. Sifry and Christopher Cerf (eds), *The Gulf War Reader: History, Documents, Opinions* (New York, 1991).
0. Bruce W. Jentleson, "The pretty prudent public: post post-Vietnam American opinion on the use of military force", *International Studies Quarterly* 36/1 (1992), quote on p. 65.
1. Michael Walzer, "Moral ambiguities in the Gulf crisis", *New Republic* (28 January 1991), p. 13.
2. Arthur Spiegelman, "Divided US waits for war, prays for peace", *Reuters News*, 15 January 1991.
3. George H. W. Bush, "Address to the nation announcing the deployment of United States armed forces to Saudi Arabia", 8 August 1990, in Peters and Woolley, *The American Presidency Project*. Available at www.presidency.ucsb.edu/ws/?pid=18750 (accessed 5 January 2014).
4. For example, US Senate Committee on Armed Services, *Crisis in the Persian Gulf Region: US Policy Options and Implications*, 101st Cong., 2nd sess., 3 December 1990, pp. 643–4 (statement of Richard B. Cheney, secretary of defense).

85. George H. W. Bush, "Remarks to Department of Defense employees", 1 August 1990, in Peters and Woolley, *The American Presidency Project*. Availabl at www.presidency.ucsb.edu/ws/?pid=18768 (accessed 5 January 2014).
86. "The Stakes in the Gulf", *Wall Street Journal*, 15 August 1990.
87. "Chief Trumpet", *New York Times*, 12 August 1990.
88. David M. Rothbard and Craig J. Rucker, "Blood and oil? Blam environmentalists", *St. Louis Dispatch*, 13 March 1991.
89. Thomas L. Friedman, "Confrontation in the Gulf", *New York Times*, 12 Augus 1990.
90. "Letters to the editor", *USA Today*, 19 January 1991.
91. Alex Molnar, "If my Marine son is killed...", *New York Times*, 23 Augus 1990.
92. Congressional Record, 1st sess., vol. 137, pt. 1 (12 January 1991), p. 98 (statement of Senator Edward M. Kennedy).
93. Out Now advertisement, published in the *Nation*, 26 November 1990, p. 645
94. Jerry Taylor, "Blood for oil?", commentary, The Cato Institute, 18 Marc 2003. Available at www.cato.org/publications/commentary/blood-oi (accessed 5 January 2014).
95. Tom Teepen, "Blood and oil", *Atlanta Journal-Constitution*, 17 January 1991
96. "14 arrested in clash outside White House", *Washington Post*, 17 January 1991
97. Figure reported in Haynes Johnson and Richard Morin, "Spoils of war" *Washington Post*, 10 March 1991.
98. Ibid.
99. "Socialists campaign for November election", *Michigan Citizen*, 3 Novembe 1990.
100. Congressional Record, 1st sess., vol. 137, pt. 1 (12 January 1991), p. 1,11 (statement of Congresswoman Marcia C. Kaptur).
101. Noam Chomsky, "Gulf War pullout", *Z Magazine*, February 1991. Availabl at www.chomsky.info/articles/199102–.htm (accessed 5 January 2014).
102. See the argument in Cyrus Bina, "The rhetoric of oil and the dilemma of wa and American hegemony", *Arab Studies Quartely* 15/3 (1993), pp. 1–20.
103. Greene, *George Bush*, p. 123.
104. Jentleson, "The pretty prudent public", p. 66.
105. Walzer, "Moral ambiguities in the Gulf crisis", p. 13.
106. "The stakes in the Gulf", *Wall Street Journal*, 15 August 1990.
107. George H. W. Bush, "Remarks at a Republican fundraising breakfast in De Moines, Iowa", 16 October 1990, in Peters and Woolley, *The America Presidency Project*. Available at www.presidency.ucsb.edu/ws/?pid=1893 (accessed 5 January 2014).
108. Bush, "The President's news conference on the Persian Gulf crisis" 8 November 1990.
109. Quote in Taylor, *War and the Media*, p. 5.
110. Congressional Record, 1st sess., vol. 137, pt. 1 (12 January 1991), p. 94 (statement of Senator James M. Jeffords).

111. Johnson and Morin, "Spoils of war."

112. See the documents collected in John Ehrenberg et al. (eds), *The Iraq Papers* (Oxford, 2010).

113. Halper and Clarke, *America Alone*, p. 4.

114. The White House, *The National Security Strategy of the United States of America*, 2002. Available at http://georgewbush-whitehouse.archives.gov/nsc/nss/2002 (accessed 6 January 2014).

115. Brewer, *Why America Fights*, pp. 271–2.

116. See, for example, a speech by Richard B. Cheney given in 2002, in which the Vice-president used a narrative of the Iraqi threat very similar to the early oil-related arguments in the run-up to the first Iraqi war in 1990. "In Cheney's words: the administration case for removing Saddam Hussein", *New York Times*, 27 August 2002.

117. Klare, *Blood and Oil*, pp. 98–9.

118. Dan Morgan and David B. Ottaway, "In Iraqi war scenario, oil is key issue", *Washington Post*, 15 September 2002. Cf. Rutledge, *Addicted to Oil*, pp. 178–81.

119. David Frum, "The curse of oil dependence", *Jerusalem Post*, 20 December 2002.

120. US Department of Defense, "Secretary Rumsfeld live interview with Infinity CBS Radio", US *Department of Defense News Transcript*, 14 November 2002. Available at www.defense.gov/Transcripts/Transcript.aspx?TranscriptID=3283 (accessed 16 July 2014).

121. Bush, "The President's news conference", 21 March 2006.

122. The document recorded a meeting about the Iraq question by a number of high-ranking British government officials, and was leaked to the press in 2005. Matthew Rycroft, "Iraq: Prime Minister's meeting, 23 July 2002 (the Downing Street memo)", in Ehrenberg, *The Iraq Papers*, p. 69.

123. US Department of Defense, "Deputy Secretary Wolfowitz interview with Sam Tannenhaus, Vanity Fair", *US Department of Defense News Transcript*, 9 May 2003. Available at www.defense.gov/transcripts/transcript.aspx?transcriptid=2594 (accessed 6 January 2014).

124. Jane K. Cramer and A. Trevor Thrall, "Why did the United States invade Iraq? A survey of International Relations and foreign policy scholars", University of Michigan, draft of a paper presented at the Annual Meeting of the International Studies Association, Chicago, Illinois (2007) p. 7. Available at www-personal.umd.umich.edu/~ atthrall/whyiraq.pdf (accessed 6 January 2014).

125. See, for example, the contributions to Jane K. Cramer and A. Trevor Thrall (eds), *Why Did the United States Invade Iraq?* (Abingdon, 2012); A. Trevor Thrall and Jane K. Cramer (eds), *American Foreign Policy and the Politics of Fear: Threat Inflation Since 9/11* (London, 2009).

126. The most frequently given reason was reshaping or influencing the Middle East. Cramer and Thrall, "Why did the United States invade Iraq? A survey", pp. 8, 37.

127. Klare, *Blood and Oil*, pp. 82–4, 94–105.

128. Rutledge, *Addicted to Oil*, ch. 12.
129. Stokes and Raphael, *Global Energy Security*, pp. 96–100.
130. Chomsky and Achcar, *Perilous Power*, p. 55.
131. Yergin, *The Quest*, p. 147.
132. Halper and Clarke, *America Alone*, pp. 13, 155.
133. Donald L. Barlett and James B. Steele, "Iraq's crude awakening", *Time*, 10 May 2003.
134. Chomsky and Achcar, *Perilous Power*, p. 55.
135. Bert Klandermans, "Peace demonstrations or antigovernment marches? The political attitudes of protestors", in Stefaan Walgrave and Dieter Rucht (eds) *The World Says No to War: Demonstrations against the War on Iraq* (Minneapolis MN, 2010), p. 100.
136. Retort, *Afflicted Powers: Capital and Spectacle in a New Age of War* (London 2005), p. 40.
137. Mary O'Brien, "Natural resistance: no hitting; protesters' signs are nuclear age-appropriate", *Eugene Weekly*, 13 February 2003. Emphasis added.
138. Representatives of these companies had attended meetings of the Future of Iraq Project.
139. Strahan, *The Last Oil Shock*, p. 16.
140. Jacob Levich, "Kick their ass and take their gas: democracy comes to Iraq" *Counterpunch*, 13 May 2003. Available at http://www.counterpunch.org/2003/05/12/democracy-comes-to-iraq (accessed 5 August 2014).
141. Retort, *Afflicted Powers*, p. 44.
142. Chinta Strausberg, "Leaders rip cost of pending war", *Chicago Defender*, 2 October 2002.
143. Chris Kutalik, "Unions lend voice to anti-war opposition", *Labor Notes*, 1 November 2002. Available at labornotes.org/node/1237 (accessed 6 January 2014).
144. Alan Greenspan, *The Age of Turbulence: Adventures in a New World* (London, 2008), p. 463.
145. Prime Minister Tony Blair on 6 February 2003. Quoted in Paul Bignell, "Secret memos expose link between oil firms and invasion of Iraq", *Independent*, 19 April 2011.
146. Chomsky and Achcar, *Perilous Power*, p. 55.
147. Cramer and Thrall, "Why did the United States invade Iraq? A survey", p. 3.
148. James E. Carter, "The State of the Union address delivered before a joint session of the Congress", 23 January 1980, in Peters and Woolley, *The American Presidency Project*. Available at www.presidency.ucsb.edu/ws/?pid=33079 (accessed 6 January 2014).
149. For example, 37 US soldiers died in a missile attack on the *USS Stark* in May 1987. See Palmer, *Guardians of the Gulf*, ch. 7.
150. Question from the audience at a public lecture. Reported in Ray McGovern, "Bush, oil, and moral bankruptcy", *Counterpunch*, 27 September 2007. Available

at www.counterpunch.org/2007/09/27/bush-oil-and-moral-bankruptcy (accessed 6 January 2014).

51. Ghosh, "Petrofiction", p. 29.

Conclusion Oil and American Identity

1. Painter, *Oil and the American Century*, p. 209.
2. Yergin, *The Prize*, p. xvi.
3. Peter Hitchcock, "Oil in an American imaginary", *New Formations* 69 (2010), p. 81.
4. Jon Stewart, *The Daily Show with Jon Stewart*, TV show, Comedy Central, 10 April 2002. Available at http://thedailyshow.cc.com/videos/pbklot/oil-pout (accessed 13 November 2011).
5. Yergin, *The Prize*, p. 598.
6. Campbell, *Writing Security*, p. 12.
7. Weldes et al., "Constructing insecurity", p. 10.
8. These figures are from the reference case scenario in tables A1, A2 and A17 in US Energy Information Administration, *Annual Energy Outlook 2014*, pp. A-1, A-3, A-33.
9. Obama, "Remarks at the Department of Energy", 5 February 2009.
10. Obama, "Remarks on energy", 26 January 2009.
11. Ian Bremmer and Kenneth A. Hersh, "When America stops importing energy", *New York Times*, 22 May 2013.
12. Barack H. Obama, "Remarks at Nashua Community College in Nashua, New Hampshire", 1 March 2012, in Peters and Woolley, *The American Presidency Project*. Available at www.presidency.ucsb.edu/ws/?pid=99751 (accessed 9 December 2013).

BIBLIOGRAPHY

Note: A small number of sources, mostly unsigned newspaper articles, are not liste in the bibliography. A full reference to these sources is given in the notes. Unles stated otherwise, newspaper articles were retrieved through the Factiva, ProQues Historical Newspapers and ProQuest News & Newspapers databases. Specific item from the ExxonMobil Historical Collection are also not listed, but a reference to th collection is given below.

All internet addresses were last accessed on 5 August 2014.

Adelman, Morris A., "Is the oil shortage real? Oil companies as OPEC tax-collectors", *Foreign Policy* (Winter 1972/73).

——— *The Genie out of the Bottle: World Oil since 1970* (Cambridge, MA, 1995).

Akins, James E., "The oil crisis: this time the wolf is here", *Foreign Affairs* 51/3 (1973), pp. 462–90.

——— "War for oil", cable from the US Embassy in Jiddah to the State Department, 13 April 1975 (National Archives and Records Administration P-Reel Index: 1975JIDDA A-23, 1975).

Alper, Donald, "Oil: the rationale for Canada's hard line", *Los Angeles Times* 7 January 1975.

Alsop, Joseph, "Oil blackmail threatens US independence", *Washington Post* 21 November 1973.

Amuzegar, Jahangir, *Managing the Oil Wealth: OPEC's Windfalls and Pitfalls* (London 2001).

Anderson, Benedict, *Imagined Communities: Reflections on the Origin and Spread o Nationalism* (London, 1991).

Anderson, Jack N., "Oil puppets", *Washington Post*, 10 June 1979.

Auerswald, Philip E., "The irrelevance of the Middle East", *American Interest* 2/5 (2007), pp. 19–37.

Auty, Richard M., *Sustaining Development in Mineral Economies: The Resource Curse Thesis* (London, 1993).

aldwin, David A., "Interdependence and power: a conceptual analysis", *International Organization* 34/4 (1980), pp. 471–506.

andow, Doug, "Befriending Saudi princes: a high price for a dubious alliance", Policy Analyis No. 428, The Cato Institute (2002). Available at www.cato.org/pubs/pas/pa-428es.html.

arca, Stefania, "Energy, property, and the Industrial Revolution narrative", *Ecological Economics* 70/7 (2011), pp. 1,309–15.

arger, Thomas C., "Suppose we seized Arab oil fields", *Los Angeles Times*, 13 April 1975.

arlett, Donald L. and James B. Steele, "Iraq's crude awakening", *Time*, 10 May 2003.

arrett, Ross and Daniel Worden, "Oil culture: guest editors' introduction", *Journal of American Studies* 46/2 (2012), pp. 269–72.

arsky, Robert B. and Lutz Kilian, "Oil and the macroeconomy since the 1970s", *Journal of Economic Perspectives* 18/4 (2004), pp. 115–34.

artelson, Jens, *A Genealogy of Sovereignty* (Cambridge, 1995).

arton, Bruce, *The Man Nobody Knows* (Indianapolis, IN, 1925).

eaubouef, Bruce A., *The Strategic Petroleum Reserve: US Energy Security and Oil Politics, 1975–2005* (College Station, TX, 2007).

erger, Peter L. and Thomas Luckmann, *The Social Construction of Reality: A Treatise in the Sociology of Knowledge* (Garden City, NY, 1966).

ignell, Paul, "Secret memos expose link between oil firms and invasion of Iraq", *Independent*, 19 April 2011.

ina, Cyrus, "The rhetoric of oil and the dilemma of war and American hegemony", *Arab Studies Quartely* 15/3 (1993), pp. 1–20.

lack, Edwin, *Internal Combustion: How Corporations and Governments Addicted the World to Oil and Derailed the Alternatives* (New York, 2006).

ohi, Douglas R., Michael A. Toman and Margaret A. Walls, *The Economics of Energy Security* (Boston, MA, 1996).

olsen, Toby and Fay Lomax Cook, "Public opinion on energy policy, 1974–2006", *Public Opinion Quarterly* 72/2 (2008), pp. 364–88.

ordetsky, Ann et al., *Securing America: Solving Our Oil Dependence Through Innovation*, National Resources Defense Council Issue Paper (2005). Available at www.nrdc.org/air/transportation/oilsecurity/plan.pdf.

remmer, Ian and Kenneth A. Hersh, "When America stops importing energy", *New York Times*, 22 May 2013.

rewer, Susan A., *Why America Fights: Patriotism and War Propaganda from the Philippines to Iraq* (Oxford, 2009).

romley, Simon, *American Hegemony and World Oil: The Industry, the State System and the World Economy* (University Park, PA, 1991).

ryce, Robert, *Gusher of Lies: The Dangerous Delusions of "Energy Independence"* (New York, 2008).

uckley, William F., "The case for the use of force to get Arab oil", *Los Angeles Times*, 10 October 1973.

——— "And what has Europe done for America lately?", *Los Angeles Times*, 15 November 1973.

uheiry, Marwan R., *US Threats of Intervention against Arab Oil, 1973–1979* (Beirut, 1980).

urns, Conrad R., "Don't buy rogue oil", *Wall Street Journal*, 8 April 2002.

Bush, George H.W., "Address to the nation announcing the deployment of Unite States armed forces to Saudi Arabia", 8 August 1990, in Gerhard Peters an John T. Woolley (eds), *The American Presidency Project, Public Papers of th Presidents*. Available at www.presidency.ucsb.edu/ws/?pid=18750.

——— "Remarks to Department of Defense employees", 15 August 1990, i Gerhard Peters and John T. Woolley (eds), *The American Presidency Projec Public Papers of the Presidents*. Available at www.presidency.ucsb.edu/ws/ pid=18768.

——— "Address before a joint session of the Congress on the Persian Gulf crisis an the federal budget deficit", 11 September 1990, in Gerhard Peters and Joh T. Woolley (eds), *The American Presidency Project, Public Papers of the President* Available at www.presidency.ucsb.edu/ws/?pid=18820.

——— "Remarks at a Republican fundraising breakfast in Des Moines, Iowa" 16 October 1990, in Gerhard Peters and John T. Woolley (eds), *The America Presidency Project, Public Papers of the Presidents*. Available at www.presidency.uc b.edu/ws/?pid=18934.

——— "The President's news conference on the Persian Gulf crisis", 8 Novembe 1990, in Gerhard Peters and John T. Woolley (eds), *The American Presidenc Project, Public Papers of the Presidents*. Available at www.presidency.ucsb.edu/ws/ pid=19019.

——— "Remarks at the Annual Convention of the National Religiou Broadcasters", 28 January 1991, in Gerhard Peters and John T. Woolley (eds *The American Presidency Project, Public Papers of the Presidents*. Available at www presidency.ucsb.edu/ws/?pid=19250.

Bush, George W., "The President's news conference with Summit of the America leaders in Quebec City", 22 April 2001, in Gerhard Peters and John T. Woolle (eds), *The American Presidency Project, Public Papers of the Presidents*. Available a www.presidency.ucsb.edu/ws/?pid=45638.

——— "Remarks on energy independence", 6 February 2003, in Gerhard Peters an John T. Woolley (eds), *The American Presidency Project, Public Papers of th Presidents*. Available at www.presidency.ucsb.edu/ws/?pid=63784.

——— "Address before a joint session of the Congress on the State of the Union" 31 January 2006, in Gerhard Peters and John T. Woolley (eds), *The America Presidency Project, Public Papers of the Presidents*. Available at www.presidency ucsb.edu/ws/?pid=65090.

——— "The President's news conference", 21 March 2006, in Gerhard Peters an John T. Woolley (eds), *The American Presidency Project, Public Papers of th Presidents*. Available at www.presidency.ucsb.edu/ws/?pid=65409.

——— "The President's news conference with European leaders in Vienna, Austria" 21 June 2006, in Gerhard Peters and John T. Woolley (eds), *The America Presidency Project, Public Papers of the Presidents*. Available at www.presidency ucsb.edu/ws/?pid=177.

——— "Address before a joint session of the Congress on the State of the Union" 23 January 2007, in Gerhard Peters and John T. Woolley (eds), *The America Presidency Project, Public Papers of the Presidents*. Available at www.presidency ucsb.edu/ws/?pid=24446.

——— "Remarks on energy in Wilmington, Delaware", 24 January 2007, i Gerhard Peters and John T. Woolley (eds), *The American Presidency Project, Publi Papers of the Presidents*. Available at www.presidency.ucsb.edu/ws/?pid=24468.

——— "Remarks on energy in Athens, Alabama", 21 June 2007, in Gerhard Peters and John T. Woolley (eds), *The American Presidency Project, Public Papers of the Presidents*. Available at www.presidency.ucsb.edu/ws/?pid=75467.

——— "Remarks during a meeting on energy security and climate change", 28 September 2007, in Gerhard Peters and John T. Woolley (eds), *The American Presidency Project, Public Papers of the Presidents*. Available at www.presidency.ucsb.edu/ws/?pid=75839.

——— "Remarks on energy", 18 June 2008, in Gerhard Peters and John T. Woolley (eds), *The American Presidency Project, Public Papers of the Presidents*. Available at www.presidency.ucsb.edu/ws/?pid=77512.

Buzan, Barry, Ole Wæver and Jaap De Wilde, *Security: A New Framework for Analysis* (Boulder, CO, 1998).

Campbell, Colin J. and Jean H. Laherrère, "The end of cheap oil", *Scientific American Magazine* (March 1998).

Campbell, David, *Writing Security: United States Foreign Policy and the Politics of Identity* (Manchester, 1992).

——— *Politics Without Principle: Sovereignty, Ethics, and the Narratives of the Gulf War* (Boulder, CO, 1993).

Carter, James E., "Address to the nation on energy", 18 April 1977, in Gerhard Peters and John T. Woolley (eds), *The American Presidency Project, Public Papers of the Presidents*. Available at www.presidency.ucsb.edu/ws/?pid=7369.

——— "The President's news conference", 28 July 1977, in Gerhard Peters and John T. Woolley (eds), *The American Presidency Project, Public Papers of the Presidents*. Available at www.presidency.ucsb.edu/ws/?pid=7878.

——— "The President's news conference", 20 July 1978, in Gerhard Peters and John T. Woolley (eds), *The American Presidency Project, Public Papers of the Presidents*. Available at www.presidency.ucsb.edu/ws/?pid=31107.

——— "National Energy Plan: address to the nation", 8 November 1978, in Gerhard Peters and John T. Woolley (eds), *The American Presidency Project, Public Papers of the Presidents*. Available at www.presidency.ucsb.edu/ws/?pid=6904.

——— "The President's news conference", 28 November 1978, in Gerhard Peters and John T. Woolley (eds), *The American Presidency Project, Public Papers of the Presidents*. Available at www.presidency.ucsb.edu/ws/?pid=31752.

——— "Address to the nation on energy and national goals: 'the Malaise Speech'", 15 July 1979, in Gerhard Peters and John T. Woolley (eds), *The American Presidency Project, Public Papers of the Presidents*. Available at www.presidency.ucsb.edu/ws/?pid=32596.

——— "The State of the Union address delivered before a joint session of the Congress", 23 January 1980, in Gerhard Peters and John T. Woolley (eds), *The American Presidency Project, Public Papers of the Presidents*. Available at www.presidency.ucsb.edu/ws/?pid=33079.

Chace, James, "American jingoism", *Harper's* (May 1976).

Cheney, Richard B., "Defending liberty in a global economy", speech at The Collateral Damage Conference, The Cato Institute, 23 June 1998. Available at www.cato.org/publications/speeches/defending-liberty-global-economy.

Chilton, Paul and George Lakoff, "Foreign policy by metaphor", in Christina Schäffner and Anita Wenden (eds), *Language and Peace* (Aldershot, 1995).

Chomsky, Noam, "Gulf War pullout", *Z Magazine*, February 1991. Available at www.chomsky.info/articles/199102–.htm.

Chomsky, Noam and Gilbert Achcar, *Perilous Power: The Middle East and US Foreign Policy: Dialogues on Terror, Democracy, War, and Justice* (Boulder, CO, 2007).
Ciută, Felix, "Conceptual notes on energy security: total or banal security", *Security Dialogue* 41/2 (2010), pp. 123–44.
Cleveland, William L., *A History of the Modern Middle East* (Boulder, CO, 2004).
Clinton, William J., "Statement on petroleum imports and energy security", 16 February 1995, in Gerhard Peters and John T. Woolley (eds), *The American Presidency Project, Public Papers of the Presidents*. Available at www.presidency.ucsb.edu/ws/?pid=50988.
CNA, *Powering America's Defense: Energy and the Risks to National Security*, report by the CNA Military Advisory Board (2009). Available at www.cna.org/reports/energy.
Cohen, Lizabeth, *A Consumers' Republic: The Politics of Mass Consumption in Postwar America* (New York, 2003).
Collier, Paul and Anke Hoeffler, "Greed and grievance in civil war", *Oxford Economic Papers* 56/4 (2004), pp. 563–95.
Conant, Melvin A., *The Oil Factor in US Foreign Policy, 1980–1990* (Lexington, KY, 1982).
Conine, Ernest, "The Kissinger 'oil-war' flap: why all the excitement?", *Los Angeles Times*, 10 January 1975.
Connolly, William E., *Identity|Difference: Democratic Negotiations of Political Paradox* (Ithaca, NY, 1991).
Cooper, Andrew S., *The Oil Kings: How the US, Iran, and Saudi Arabia Changed the Balance of Power in the Middle East* (New York, 2011).
Copeland, Dale C., "Economic interdependence and war: a theory of trade expectations", *International Security* 20/4 (1996), pp. 5–41.
Cox, Robert W., "Social forces, states and world orders: beyond international relations theory", *Millennium: Journal of International Studies* 10/2 (1986), pp. 126–55.
Cramer, Jane K. and A. Trevor Thrall, "Why did the United States invade Iraq? A survey of International Relations and foreign policy scholars", University of Michigan, draft of a paper presented at the Annual Meeting of the International Studies Association, Chicago, Illinois (2007). Available at www-personal.umd.umich.edu/~atthrall/whyiraq.pdf.
——— (eds), *Why Did the United States Invade Iraq?* (Abingdon, 2012).
Crane, Keith et al., *Imported Oil and US National Security* (Santa Monica, CA 2009).
Daniel, Clifton, "Kissinger remark on force sparks wide speculation", *New York Times*, 7 January 1975.
Davenport, Ernest H. and Sidney R. Cooke, *The Oil Trusts and Anglo-American Relations* (New York, 1924).
Doty, Roxanne Lynn, "Foreign policy as social construction: a post-positivist analysis of US counterinsurgency policy in the Philippines", *International Studies Quarterly* 37/3 (1993), pp. 297–320.
Dreyfuss, Robert, "The thirty-year itch", *Mother Jones* (March/April 2003).
Duffield, John S., *Over a Barrel: The Costs of US Foreign Oil Dependence* (Stanford, CA, 2008).
Dyke, Willard Van, *American Frontier*, short film by Affiliated Film Producers and the American Petroleum Institute (1953). Available at www.archive.org/details/American1953.
Ehrenberg, John et al. (eds), *The Iraq Papers* (Oxford, 2010).

hrlich, Paul R., *The Population Bomb* (New York, 1971).

isenhower, Dwight D., "Annual message to the Congress on the State of the Union", 10 January 1957, in Gerhard Peters and John T. Woolley (eds), *The American Presidency Project, Public Papers of the Presidents*. Available at www.presidency.ucsb.edu/ws/?pid=11029.

nders, Thomas O., "OPEC and the industrial countries: the next ten years", *Foreign Affairs* 53/4 (1975), pp. 625–37.

pstein, Charlotte, *The Power of Words in International Relations: Birth of an Anti-Whaling Discourse* (Cambridge, MA, 2008).

sposito, John L., *The Islamic Threat: Myth or Reality?* (New York, 1999).

arhar, Barbara C., "Public opinion about energy", *Public Opinion Quarterly* 58/4 (1994), pp. 603–32.

eatherstone, Mike, *Consumer Culture and Postmodernism* (Los Angeles, 2007).

ischer, David H., *Liberty and Freedom: A Visual History of America's Founding Ideas* (Oxford, 2005).

ogel, Cathleen, "Constructing progressive climate change norms: the US in the early 2000s", in Mary E. Pettenger (ed.), *The Social Construction of Climate Change: Power, Knowledge, Norms, Discourses* (Aldershot, 2007).

ord, Gerald R., "Remarks to the Ninth World Energy Conference, Detroit, Michigan", 23 September 1974, in Gerhard Peters and John T. Woolley (eds), *The American Presidency Project, Public Papers of the Presidents*. Available at www.presidency.ucsb.edu/ws/?pid=4732.

——— "Address to a joint session of the Congress on the economy", 8 October 1974, in Gerhard Peters and John T. Woolley (eds), *The American Presidency Project, Public Papers of the Presidents*. Available at www.presidency.ucsb.edu/ws/?pid=4434.

——— "Address to the nation on energy and economic programs", 13 January 1975, in Gerhard Peters and John T. Woolley (eds), *The American Presidency Project, Public Papers of the Presidents*. Available at www.presidency.ucsb.edu/ws/?pid=4916.

——— "Address before a joint session of the Congress reporting on the State of the Union", 15 January 1975, in Gerhard Peters and John T. Woolley (eds), *The American Presidency Project, Public Papers of the Presidents*. Available at www.presidency.ucsb.edu/ws/?pid=4938.

——— "The President's news conference", 21 January 1975, in Gerhard Peters and John T. Woolley (eds), *The American Presidency Project, Public Papers of the Presidents*. Available at www.presidency.ucsb.edu/ws/?pid=5027.

——— "Address to the nation on energy programs", 27 May 1975, in Gerhard Peters and John T. Woolley (eds), *The American Presidency Project, Public Papers of the Presidents*. Available at www.presidency.ucsb.edu/ws/?pid=4942.

——— "Special message to the Congress on energy", 7 January 1977, in Gerhard Peters and John T. Woolley (eds), *The American Presidency Project, Public Papers of the Presidents*. Available at www.presidency.ucsb.edu/ws/?pid=5542.

oucault, Michel, *Discipline and Punish: The Birth of the Prison* (London, 1977).

riedman, Thomas L., "Confrontation in the Gulf", *New York Times*, 12 August 1990.

——— "Will pigs fly?", *New York Times*, 3 February 2006.

——— "No fooling mother nature", *New York Times*, 4 May 2010.

ritchey, Clayton, "'Project Independence': freedom from OPEC", *Washington Post*, 11 June 1979.

Frum, David, "The curse of oil dependence", *Jerusalem Post*, 20 December 2002.

Gallup, "Americans rate computer industry best, oil and gas worst", 16 August 2012. Available at www.gallup.com/poll/156713/Americans-Rate-Computer-Industry-Best-Oil-Gas-Worst.aspx.

Ghosh, Amitav, "Petrofiction: the oil encounter and the novel", *New Republic* (2 March 1992).

Goldemberg, José and Oswaldo Lucon, *Energy, Environment and Development* (London, 2010).

Goodstein, David L., *Out of Gas: The End of the Age of Oil* (New York, 2004).

Gore, Albert A., "Still hooked on oil", *New York Times*, 6 March 1991.

Gottdiener, Mark, "Approaches to consumption: classical and contemporary perspectives", in Mark Gottdiener (ed.), *New Forms of Consumption: Consumers, Culture, and Commodification* (Lanham, MD, 2000).

Grant, Linda and Robert A. Rosenblatt, "Gas rationing scare: where do we stand?" *Los Angeles Times*, 8 December 1978.

Greene, John R., *The Presidency of George Bush* (Lawrence, KS, 2000).

Greenspan, Alan, *The Age of Turbulence: Adventures in a New World* (London, 2008).

Hall, Stuart, "Signification, representation, ideology: Althusser and the post-structuralist debates", *Critical Studies in Mass Communication* 2/2 (1985), pp. 91–114.

Hallett, Steve and John Wright, *Life Without Oil: Why We Must Shift to a New Energy Future* (Amherst, NY 2011).

Halper, Stefan A. and Jonathan Clarke, *America Alone: The Neo-Conservatives and the Global Order* (Cambridge, 2004).

———, *The Silence of the Rational Center: Why American Foreign Policy is Failing* (New York, 2007).

Hansen, Lene, *Security as Practice: Discourse Analysis and the Bosnian War* (London, 2006).

Harlow, Rachel Martin, "Agency and agent in George Bush's Gulf War rhetoric", in Martin J. Medhurst (ed.), *The Rhetorical Presidency of George H.W. Bush* (College Station, TX, 2006).

Hartshorn, J.E., *Oil Trade: Politics and Prospects* (Cambridge, 1993).

Harvey, Arnold D., *Body Politic: Political Metaphor and Political Violence* (Newcastle-upon-Tyne, 2007).

Head, Simon, "The monarchs of the Persian Gulf", *The New York Review of Books* (21 March 1974).

Heinberg, Richard, *The Party's Over: Oil, War and the Fate of Industrial Societies* (Gabriola Island, BC, 2005).

Hickel, Walter J., "The energy war II: what we must do at home", *Reader's Digest* (February 1974).

Higgins, Andrew, "Power and peril: America's supremacy and its limits", *Wall Street Journal*, 4 February 2004.

Hitchcock, Peter, "Oil in an American imaginary", *New Formations* 69 (2010), pp. 81–97.

Hodel, Donald P., "National Energy Policy Plan: a preview", *Vital Speeches of the Day* 50/5 (1983), pp. 130–3.

Hoffmann, Stanley, "Letter to the editor", *Commentary* (April 1975).

Hofmeister, John, *Why We Hate the Oil Companies: Straight Talk from an Energy Insider* (New York, 2010).

Iorowitz, Daniel, *The Anxieties of Affluence: Critiques of American Consumer Culture, 1939–1979* (Amherst, NY, 2004).

Iughes, Thomas P., "Technological momentum", in Merrit Roe Smith and Leo Marx (eds), *Does Technology Drive History?: The Dilemma of Technological Determinism* (Cambridge, MA, 1994).

Iunt, Michael H., *Ideology and US Foreign Policy* (New Haven, CT, 2009).

Iuntington, Samuel P., "The clash of civilizations?", *Foreign Affairs* 72/3 (1993), pp. 22–49.

:kes, Harold L., "After the deluge: what price gasoline?", *Saturday Evening Post* (16 February 1935).

gnotus, Miles, "Seizing Arab oil", *Harper's* (March 1975).

kenberry, G. John, *After Victory: Institutions, Strategic Restraint, and the Rebuilding of Order after Major Wars* (Princeton, NJ, 2001).

nstitute of Labor and Industrial Relations, *Contribution of the Automotive Industry to the US Economy in 1998*, a study prepared for the Alliance of Automobile Manufacturers and the Association of International Automobile Manufacturers (2001). Available at http://deepblue.lib.umich.edu/bitstream/2027.42/1483/2/94310.0001.001.pdf.

vins, Molly, "120 tribes form new group to block Anti-Indian moves", *New York Times*, 16 April 1978.

acobs, Matthew F., *Imagining the Middle East: The Building of an American Foreign Policy, 1918–1967* (Chapel Hill, NC, 2011).

entleson, Bruce W., "The pretty prudent public: post post-Vietnam American opinion on the use of military force", *International Studies Quarterly* 36/1 (1992), pp. 49–73.

offé, George, "Confrontational mutual perceptions and images: orientalism and occidentalism in Europe and the Islamic world", *The International Spectator: Italian Journal of International Affairs* 42/2 (2007), pp. 161–77.

offé, George et al., "Expropriation of oil and gas investments: historical, legal and economic perspectives in a new age of resource nationalism", *Journal of World Energy Law & Business* 2/1 (2009), pp. 3–23.

ohnson, Haynes and Richard Morin, "Spoils of war", *Washington Post*, 10 March 1991.

Kamalipour, Yahya R. (ed.), *The US Media and the Middle East: Image and Perception* (Westport, CT, 1995).

Karpel, Craig S., "Ten ways to break OPEC", *Harper's* (January 1976).

Kempster, Norman and Ronald J. Ostrow, "Oil: the tie that binds US, Arabs", *Los Angeles Times*, 22 November 1978.

Kennedy, Donald, "Beyond petroleum?", *Science* 329/5993 (2010), p. 727.

Kenyon, Curtis, *Tulsa*, motion picture by Walter Wanger Productions (1949). Available at www.youtube.com/watch?v=WO7WPxZt63w.

Keohane, Robert O., *After Hegemony: Cooperation and Discord in the World Political Economy* (Princeton, NJ, 2005).

Keohane, Robert O. and Joseph S. Nye, *Power and Interdependence* (New York, 2001).

Kissinger, Henry A., "The Washington Energy Conference: interdependence and cooperative solutions, 11 February 1974", *Vital Speeches of the Day* 40/11 (1974), pp. 322–5.

Klandermans, Bert, "Peace demonstrations or antigovernment marches? The political attitudes of protestors", in Stefaan Walgrave and Dieter Rucht (eds),

The World Says No to War: Demonstrations against the War on Iraq (Minneapoli MN, 2010).

Klare, Michael T., *Resource Wars: The New Landscape of Global Conflict* (New York 2001).

——— *Blood and Oil: The Dangers and Consequences of America's Growing Petroleum Dependency* (New York, 2004).

——— "There will be blood: political violence, regional warfare, and the risk of great-power conflict over contested energy sources", in Gal Luft and Anne Kori (eds), *Energy Security Challenges for the 21st Century: A Reference Handbook* (Sant Barbara, CA, 2009).

Krasner, Stephen D., *Structural Conflict: The Third World against Global Liberalism* (Berkeley, CA, 1985).

Krauss, Clifford and Eric Lipton, "US inches toward goal of energy independence" *New York Times*, 22 March 2012.

Krauthammer, Charles, "A nation of oil addicts", *Washington Post*, 17 March 200

——— "Energy independence? A serious plan requires taxes, ANWR and nukes" *Washington Post*, 26 January 2007.

Kutalik, Chris, "Unions lend voice to anti-war opposition", *Labor Notes*, 1 Novembe 2002. Available at labornotes.org/node/1237.

LaDuke, Winona, *The Winona LaDuke Reader: A Collection of Essential Writing* (Stillwater, MN, 2002).

Lakoff, George and Mark Johnson, *Metaphors We Live By* (Chicago, 1980).

Laqueur, Walter and Edward Luttwak, "Oil", *Commentary* (October 1973).

Latham, Robert, *The Liberal Moment: Modernity, Security, and the Making of Postwa International Order* (New York, 1997).

Levi, Michael, "Think again: the American energy boom", *Foreign Policy* (July August 2012).

Levich, Jacob, "Kick their ass and take their gas: democracy comes to Iraq" *Counterpunch*, 13 May 2003. Available at http://www.counterpunch.org/2003 05/12/democracy-comes-to-iraq.

Levy, Walter J., "Oil and the Decline of the West", *Foreign Affairs* 58/5 (1980) pp. 999–1,015.

Lewis, Bernard, "The roots of Muslim rage", *Atlantic Monthly* (September 1990).

Lipset, Seymour M., "Some social requisites of democracy: economic developmen and political legitimacy", *American Political Science Review* 53/1 (1959) pp. 69–105.

Little, Douglas, *American Orientialism: The United States and the Middle East since 194* (Chapel Hill, NC, 2008).

Lovins, Amory B., "Energy strategy: the road not taken?", *Foreign Affairs* 55/ (1976), pp. 65–96.

Luciani, Giacomo, "Oil and political economy in the international relations of th Middle East", in Louise Fawcett (ed.), *International Relations of the Middle Eas* (Oxford, 2009).

Lutz, Catherine A. and Jane L. Collins, *Reading National Geographic* (Chicago, 1993)

Luft, Gal and Anne Korin, "The myth of US energy dependence", *Foreign Affairs* 15 October 2013. Available at www.foreignaffairs.com/articles/140172/gal luft-and-anne-korin/the-myth-of-us-energy-dependence.

MacCurdy, George G., *Human Origins: A Manual of Prehistory*, vol. II (New York 1933).

Machan, Tibor R., "Letter to the editor", *Commentary* (April 1975).

Maugeri, Leonardo, "Not in oil's name", *Foreign Affairs* 82/4 (2003), pp. 165–74.

——— *The Age of Oil: The Mythology, History, and Future of the World's Most Controversial Resource* (Westport, CT, 2006).

McAlister, Melani, *Epic Encounters: Culture, Media, and US Interests in the Middle East, 1945–2000* (Berkeley, CA, 2001).

McClintock, Anne, *Imperial Leather: Race, Gender and Sexuality in the Colonial Contest* (New York, 1995).

McCormick, Thomas J., *America's Half-Century: United States Foreign Policy in the Cold War and After* (Baltimore, MD, 1995).

McGovern, Ray, "Bush, oil, and moral bankruptcy", *Counterpunch*, 27 September 2007. Available at www.counterpunch.org/2007/09/27/bush-oil-and-moral-bankruptcy.

McMillan, Susan M., "Interdependence and conflict", *Mershon International Studies Review* 41/1 (1997), pp. 33–58.

Meadows, Donella et al., *The Limits to Growth: A Report for the Club of Rome's Project on the Predicament of Mankind* (London, 1972).

Mearsheimer, John J., "Disorder restored", in Graham Allison and Gregory F. Treverton (eds), *Rethinking America's Security: Beyond Cold War to New World Order* (New York, 1992).

Melby, Eric D.K., *Oil and the International System: The Case of France, 1918–1969* (New York, 1981).

Melosi, Martin, *Coping with Abundance: Energy and Environment in Industrial America* (Philadelphia, PA, 1985).

——— "Energy transitions in historical perspective", in Brendan M. Dooley (ed.), *Energy and Culture: Perspectives on the Power to Work* (Aldershot, 2006).

Merrill, Karen R., *The Oil Crisis of 1973–1974: A Brief History with Documents* (Boston, MA, 2007).

Milloy, Courtland, "Driven to war on a full tank of questions", *Washington Post*, 21 October 2001.

Mitchell, Timothy, *Carbon Democracy: Political Power in the Age of Oil* (London, 2011).

Molnar, Alex, "If my Marine son is killed...", *New York Times*, 23 August 1990.

Morgan, Dan and David B. Ottaway, "In Iraqi war scenario, oil is key issue", *Washington Post*, 15 September 2002.

Morgenthau, Hans J. and Kenneth W. Thompson, *Politics Among Nations: The Struggle for Power and Peace* (New York, 1985).

Morse, Edward L., *Energy 2020: Independence Day*, Citi GPS: Global Perspectives & Solutions (2013). Available at https://www.citivelocity.com/citigps/ReportSeries.action?recordId=16.

Morse, Edward L. and Amy Myers Jaffe, *Strategic Energy Policy Challenges of the 21st Century: Report of An Independent Task Force*, cosponsored by the James A. Baker III Institute for Public Policy of Rice University and the Council on Foreign Relations (2001). Available at www.cfr.org/energy-policy/strategic-energy-policy-challenges-21st-century/p3942.

Mouawad, Jad and Matthew L. Wald, "The oil uproar that isn't", *New York Times*, 12 July 2005.

Mullaney, Tim, "Energy independence isn't just a pipe dream", *USA Today*, 16 May 2012.

Munif, Abdelrahman, *Cities of Salt* (New York, 1987).

Nanos, Nik, "Entering the energy & environment policy frontier: an examination of the intersection of public opinion and public policy", Woodrow Wilson International Center for Scholars, 26 August 2013. Available at www.wilsoncenter.org/article/new-report-entering-the-energy-environment-policy-frontier.

The National Bureau of Economic Research (NBER), *US Business Cycle Expansions and Contractions* (2013). Available at www.nber.org/cycles/cyclesmain.html.

Nivola, Pietro S., "Energy independence or interdependence? Integrating the North American energy market", *The Brookings Review* 20/2 (2002), pp. 24–7.

Nixon, Richard M., "Special message to the Congress on energy resources", 4 June 1971, in Gerhard Peters and John T. Woolley (eds), *The American Presidency Project, Public Papers of the Presidents*. Available at www.presidency.ucsb.edu/ws/?pid=3038.

——— "The President's news conference", 5 September 1973, in Gerhard Peters and John T. Woolley (eds), *The American Presidency Project, Public Papers of the Presidents*. Available at www.presidency.ucsb.edu/ws/?pid=3948.

——— "Address to the nation about policies to deal with the energy shortages", 7 November 1973, in Gerhard Peters and John T. Woolley (eds), *The American Presidency Project, Public Papers of the Presidents*. Available at www.presidency.ucsb.edu/ws/?pid=4034.

——— "Address to the nation about national energy policy", 25 November 1973, in Gerhard Peters and John T. Woolley (eds), *The American Presidency Project, Public Papers of the Presidents*. Available at www.presidency.ucsb.edu/ws/?pid=4051.

——— "Radio address about the national energy crisis", 19 January 1974, in Gerhard Peters and John T. Woolley (eds), *The American Presidency Project, Public Papers of the Presidents*. Available at www.presidency.ucsb.edu/ws/?pid=4208.

——— "Special message to the Congress on the energy crisis", 23 January 1974, in Gerhard Peters and John T. Woolley (eds), *The American Presidency Project, Public Papers of the Presidents*. Available at www.presidency.ucsb.edu/ws/?pid=4241.

——— "Address on the State of the Union delivered before a joint session of the Congress", 30 January 1974, in Gerhard Peters and John T. Woolley (eds), *The American Presidency Project, Public Papers of the Presidents*. Available at www.presidency.ucsb.edu/ws/?pid=4327.

Noreng, Øystein, *Oil and Islam: Social and Economic Issues* (Chichester, 1997).

——— "Islam and oil", in Eni S.P.A (Ente Nazionale Idrocarburi) (ed.), *Encyclopedia of Hydrocarbons (vol. IV: Economics, Policies, and Legislation)* (Rome, 2006).

Nye, David E., *Image Worlds: Corporate Identities at General Electric, 1890–1930* (Cambridge, MA, 1985).

——— "Energy narratives", *American Studies in Scandinavia* 25/2 (1993), pp. 73–91.

——— *Consuming Power: A Social History of American Energies* (Cambridge, MA, 2001).

Nye, Joseph S., "Energy nightmares", *Foreign Policy* (Autumn 1980).

——— *Understanding International Conflicts: An Introduction to Theory and History* (New York, 2009).

O'Brien, Mary, "Natural resistance: no hitting – protesters' signs are nuclear age-appropriate", *Eugene Weekly* (13 February 2003).

Obama, Barack H., "Remarks on energy", 26 January 2009, in Gerhard Peters and John T. Woolley (eds), *The American Presidency Project, Public Papers of the Presidents*. Available at www.presidency.ucsb.edu/ws/?pid=85689.

——— "Remarks at the Department of Energy", 5 February 2009, in Gerhard Peters and John T. Woolley (eds), *The American Presidency Project, Public Papers of the Presidents*. Available at www.presidency.ucsb.edu/ws/?pid=85735.

——— "The President's weekly address", 7 February 2009, in Gerhard Peters and John T. Woolley (eds), *The American Presidency Project, Public Papers of the Presidents*. Available at www.presidency.ucsb.edu/ws/?pid=85738.

——— "The President's weekly address", 4 July 2009, in Gerhard Peters and John T. Woolley (eds), *The American Presidency Project, Public Papers of the Presidents*. Available at www.presidency.ucsb.edu/ws/?pid=86356.

——— "Address to the nation on the oil spill in the Gulf of Mexico", 15 June 2010, in Gerhard Peters and John T. Woolley (eds), *The American Presidency Project, Public Papers of the Presidents*. Available at www.presidency.ucsb.edu/ws/?pid=88033.

——— "Remarks at Nashua Community College in Nashua, New Hampshire", 1 March 2012, in Gerhard Peters and John T. Woolley (eds), *The American Presidency Project, Public Papers of the Presidents*. Available at www.presidency.ucsb.edu/ws/?pid=99751.

Olien, Roger M. and Diana Davids Olien, *Oil and Ideology: The Cultural Creation of the American Petroleum Industry* (Chapel Hill, NC, 2000).

Oren, Michael B., *Power, Faith and Fantasy: America in the Middle East, 1776 to the Present* (New York, 2007).

Ó Tuathail, Gearóid and John Agnew, "Geopolitics and discourse: practical geopolitical reasoning in American foreign policy", *Political Geography* 11/2 (1992), pp. 190–204.

Painter, David S., *Oil and the American Century: The Political Economy of US Foreign Oil Policy, 1941–1954* (Baltimore, MD, 1986).

——— "International oil and national security", *Daedalus* 120/4 (1991), pp. 183–206.

——— "Oil", in Alexander DeConde, Fredrik Logevall and Richard Dean Burns (eds), *Encyclopedia of American Foreign Policy*, vol. 3 (New York, 2002).

——— "The Marshall Plan and oil", *Cold War History* 9/2 (2009), pp. 159–75.

Palmer, Michael A., *Guardians of the Gulf: A History of America's Expanding Role in the Persian Gulf, 1833–1992* (New York, 1992).

Parra, Francisco, *Oil Politics: A Modern History of Petroleum* (London, 2010).

Parry, Ian W.H. and Joel Darmstadter, "The costs of US oil dependency", *Resources for the Future*, Discussion Paper 03–59 (2003). Available at www.rff.org/documents/rff-dp-03–59.pdf.

Perelman, Lewis J., "Speculations on the transition to sustainable energy", *Ethics* 90/3 (1980), pp. 392–416.

Plummer, James L. (ed.), *Energy Vulnerability* (Cambridge, MA, 1982).

Potter, David M., *People of Plenty: Economic Abundance and the American Character* (Chicago, 1954).

Randall, Stephen J., *United States Foreign Oil Policy Since World War I: For Profits and Security* (Montreal, 2005).

Rashid, Ahmed, *Taliban: Militant Islam, Oil and Fundamentalism in Central Asia* (London, 2010).

Raspberry, William, "Blackmail—or diplomacy?", *Washington Post*, 23 November 1973.
Reagan, Ronald W., "Message to the Congress transmitting the National Energy Policy Plan", 17 July 1981, in Gerhard Peters and John T. Woolley (eds) *The American Presidency Project, Public Papers of the Presidents*. Available at www.presidency.ucsb.edu/ws/?pid=44096.
——— "Address before a joint session of Congress on the State of the Union", 4 February 1986, in Gerhard Peters and John T. Woolley (eds), *The American Presidency Project, Public Papers of the Presidents*. Available at www.presidency.ucsb.edu/ws/?pid=36646.
——— "Radio address to the nation on oil prices", 19 April 1986, in Gerhard Peters and John T. Woolley (eds), *The American Presidency Project, Public Papers of the Presidents*. Available at www.presidency.ucsb.edu/ws/?pid=37156.
Reed, Stanley and Alison Fitzgerald, *In Too Deep: BP and the Drilling Race That Took It Down* (Hoboken, NJ, 2011).
Retort, *Afflicted Powers: Capital and Spectacle in a New Age of War* (London, 2005).
Roberts, Paul, *The End of Oil: On the Edge of a Perilous New World* (Boston, 2005).
Rodgers, Daniel T., "Exceptionalism", in Anthony Mohlo and Gordon S. Wood (eds), *Imagined Histories: American Historians Interpret the Past* (Princeton, NJ 1998).
Rosecrance, Richard, *The Rise of the Trading State: Commerce and Conquest in the Modern World* (New York, 1986).
Rosecrance, Richard and Arthur Stein, "Interdependence: myth or reality?", *World Politics* 24/1 (1973), pp. 1–27.
Rosenberg, Emily S., *Spreading the American Dream: American Economic and Cultural Expansion, 1890–1945* (New York, 1982).
Rosenfeld, Stephen S., "Oil and military force", *Washington Post*, 10 January 1975.
Ross, Michael L., "The political economy of the resource curse", *World Politics* 51/2 (1999), pp. 297–322.
Rosser, Andrew, "The political economy of the resource curse: a literature survey", Working Paper No. 268, Institute of Development Studies (IDS), Centre for the Future State (2006). Available at www.ids.ac.uk/files/WP268.pdf.
Rothbard, David M. and Craig J. Rucker, "Blood and oil? Blame environmentalists", *St. Louis Dispatch*, 13 March 1991.
Rowen, Hobart, "The 'economic agression' of the Arabs", *Washington Post*, 6 December 1973.
——— "A do-nothing oil policy", *Washington Post*, 22 February 1979.
——— "OPEC's money grab", *Washington Post*, 19 April 1979.
Rudolph, Barbara, "Step on the gas, pay the price", *Time*, 3 April 1989.
Russett, Bruce and Miroslav Nincic, "American opinion on the use of military force abroad", *Political Science Quarterly* 91/3 (1976), pp. 411–31.
Rutledge, Ian, *Addicted to Oil: America's Relentless Drive for Energy Security* (London, 2006).
Rycroft, Matthew, "Iraq: Prime Minister's meeting, 23 July 2002 (the Downing Street memo)", in John Ehrenberg et al. (eds), *The Iraq Papers* (Oxford, 2010).
Said, Edward W., *Culture and Imperialism* (New York, 1994).
——— *Covering Islam: How the Media and the Experts Determine How We See the Rest of the World* (New York, 1997).
——— *Orientalism* (London, 2003).

amuelson, Robert J., "Now do we get serious on oil?", *Washington Post*, 11 October 2001.

chultze, Charles L., "Oil war: fantasy no one needs", *Washington Post*, 26 January 1975.

chumacher, Ernst F., *Small is Beautiful: A Study of Economics as if People Mattered* (London, 1973).

churr, Sam H. and Bruce C. Netschert, *Energy in the American Economy, 1850–1975* (Baltimore, MD, 1960).

ecuring America's Future Energy (SAFE), *Oil Dependence: A Threat to US Economic & National Security*, Policy Report (2006). Available at www.secureenergy.org/sites/default/files/155_Briefing-OilDependence.pdf.

eib, Gerald F., "How oil dependence fuels US policies", *Wall Street Journal*, 22 August 2005.

haheen, Jack G., *Reel Bad Arabs: How Hollywood Vilifies a People* (Northampton, MA, 2001).

——— *Guilty: Hollywood's Verdict on Arabs after 9/11* (Northampton, MA, 2008).

hapiro, Michael J., "Moral geographies and the ethics of post-sovereignty", *Public Culture* 6 (1994), pp. 479–502.

ifry, Micah L. and Christopher Cerf (eds), *The Gulf War Reader: History, Documents, Opinions* (New York, 1991).

impson, George E. and J. Milton Yinger, *Racial and Cultural Minorities: An Analysis of Prejudice and Discrimination* (New York, 1985).

inclair, Upton, *Oil!* (London, 2008).

keet, Ian, *OPEC: Twenty-Five Years of Prices and Politics* (Cambridge, 1988).

mith, Frederick W. and Paul X. Kelley, "Are we ready for the next oil shock?", *Washington Post*, 11 August 2006.

piegelman, Arthur, "Divided US waits for war, prays for peace", *Reuters News*, 15 January 1991.

t. Clair, David J., *The Motorization of American Cities* (New York, 1986).

telzer, Irwin M., "Can we do without Saudi oil?", *Weekly Standard*, 19 November 2001. Available at www.weeklystandard.com/Content/Public/Articles/000/000/000/525pggsa.asp.

tephanson, Anders, *Manifest Destiny: American Expansionism and the Empire of Right* (New York, 1996).

tevens, Paul, "National oil companies and international oil companies in the Middle East: under the shadow of government and the resource nationalism cycle", *Journal of World Energy Law & Business* 1/1 (2008), pp. 5–30.

——— "The 'shale gas revolution': developments and changes", Briefing Paper EERG BP 2012/04, Chatham House (2012). Available at www.chathamhouse.org/publications/papers/view/185311.

tevens, Paul and Evelyn Dietsche, "Resource curse: an analysis of causes, experiences and possible ways forward", *Energy Policy* 36/1 (2008), pp. 56–65.

tobaugh, Robert and Daniel Yergin, "After the second shock: pragmatic energy strategies", *Foreign Affairs* 57/4 (1979), pp. 836–71.

tokes, Doug and Sam Raphael, *Global Energy Security and American Hegemony* (Baltimore, MD, 2010).

tone, Isidor F., "War for oil?", *New York Review of Books* (6 February 1975).

trahan, David, *The Last Oil Shock: A Survival Guide to the Imminent Extinction of Petroleum Man* (London, 2008).

Strausberg, Chinta, "Leaders rip cost of pending war", *Chicago Defender*, 2 Octobe 2002.

Szeman, Imre, "System failure: oil, futurity, and the anticipation of disaster", *Sout Atlantic Quarterly* 106/4 (2007), pp. 805–23.

Taylor, Jerry, "Blood for oil?", commentary, The Cato Institute, 18 March 200 Available at www.cato.org/publications/commentary/blood-oil.

Taylor, Jerry and Peter Van Doren, "The energy security obsession", *Georgetow Journal of Law and Public Policy* 6/2 (2008), pp. 475–86.

Taylor, Philip M., *War and the Media: Propaganda and Persuasion in the Gulf Wa* (Manchester, 1998).

Teepen, Tom, "Blood and oil", *Atlanta Journal-Constitution*, 17 January 1991.

Thrall, A. Trevor and Jane K. Cramer (eds), *American Foreign Policy and the Politics Fear: Threat Inflation Since 9/11* (London, 2009).

Tucker, Robert W., "Oil: the issue of American intervention", *Commentary* (Januar 1975).

——— "Further reflections on oil & force", *Commentary* (March 1975).

——— "Letter to the editor", *Commentary* (April 1975).

Tyrrell, R. Emmett, "Chimera in the Middle East", *Harper's* (November 1976).

United Nations General Assembly, "A/RES/S-6/3201, resolution adopted by th General Assembly 3201 (S-VI): declaration on the establishment of a ne international economic order", 1 May 1974. Available at www.un-document net/s6r3201.htm.

Urbano, Carl, *Destination Earth*, animated short film by Sutherland Productions an the American Petroleum Institute (1956). Available at http://archive.org details/Destinat1956.

Urry, John, "The 'system' of automobility", *Theory, Culture & Society* 21/4 (2004 pp. 25–39.

US Cabinet Task Force on Oil Import Control, *The Oil Import Question: A Report on th Relationship of Oil Imports to the National Security* (Washington, DC, 1970).

US Congress, Congressional Record, 93rd Cong., 1st sess., vol. 119, pt. 1 (Washington, DC, 1973).

——— 94th Cong., 1st sess., vol. 121, pt. 1 (Washington, DC, 1975).

——— 102nd Cong., 1st sess., vol. 137, pt. 1 (Washington, DC, 1991).

US Congress, House, Committee on International Relations, Special Subcommitte on Investigations, *Oil Fields as Military Objectives: A Feasibility Study*, repor prepared by the Congressional Research Service, 94th Cong., 1st sess. Committee Print 56 (Washington, DC, 1975).

US Congress, House, Committee on Ways and Means, *Hearings on the Energy Crisi and Proposed Solutions*, 94th Cong., 1st sess., 3–17 March 1975.

US Congress, Senate, Committee on Armed Services, *Crisis in the Persian Gulf Region US Policy Options and Implications*, 101st Cong., 2nd sess., 3 December 1990.

US Congress, Senate, Committee on Banking, Housing and Urban Affairs, *Hearin on S. 2532 to Establish the Energy Independence Authority*, 94th Cong., 2nd sess., 12 14 and 18 April and 10 May 1976.

US Congress, Senate, Committee on Energy and Natural Resources, *Hearing o Energy Independence*, 109th Cong., 2nd sess., 7 March 2006.

——— *Hearing on the Geopolitics of Oil and Its Implications for US Economic an National Security*, 110th Cong., 1st sess., 10 January 2007.

US Congress, Senate, Committee on Energy and Natural Resources, Subcommittee on Energy Regulation, *Hearing on S. 1134, S. 1205, S. 1417 and S. 1470 to Limit Oil Imports*, 96th Cong., 1st sess., 18 September 1979.

US Congress, Senate, Committee on Finance, *Bills to Encourage National Industrial Recovery, to Foster Fair Competition, and to Provide for the Construction of Certain Useful Public Works, and for Other Purposes, Hearings on S. 1712 and H.R. 5755*, 73rd Cong., 1st sess., 18–20 May 1933.

US Congress, Senate, Committee on Foreign Relations, *Hearing on the Implications of the Current Energy Problem for United States Foreign Policy*, 93rd Cong., 1st sess., 30 and 31 May 1973.

——— *Hearing on Energy Security and Oil Dependence*, 109th Cong., 2nd sess., 16 May 2006.

——— *Hearing on Energy Security: Historical Perspectives and Modern Challenges*, 111th Cong., 1st sess., 12 May 2009.

US Congress, Senate, Committee on Foreign Relations, Subcommittee on International Economic Policy, Export and Trade Promotion, *Hearing on Overview of Global Energy Security Issues*, 108th Cong., 1st sess., 8 April 2003.

US Congress, Senate, Committee on Foreign Relations, Subcommittee on Near Eastern and South Asian Affairs, *Hearings on Middle East Peace Prospects*, 94th Cong., 2nd sess., 19 May, 7, 21, 30 June, 19 and 26 July 1976.

——— *Hearing on US Foreign Policy, Petroleum, and the Middle East*, 109th Cong., 1st sess., 20 October 2005.

US Congress, Senate, Committee on Interior and Insular Affairs, *Hearing on the President's Energy Message and S. 1570*, 93rd Cong., 1st sess., 1 May 1973.

US Department of Defense, "Secretary Rumsfeld live interview with Infinity CBS Radio", *US Department of Defense News Transcript*, 14 November 2002. Available at www.defense.gov/Transcripts/Transcript.aspx?TranscriptID=3283.

——— "Deputy Secretary Wolfowitz interview with Sam Tannenhaus, Vanity Fair", *US Department of Defense News Transcript*, 9 May 2003. Available at www.defense.gov/transcripts/transcript.aspx?transcriptid=2594.

US Department of State, *Foreign Relations of the United States (FRUS): 1945–1953, Harry S Truman, Volume VIII (The Near East and Africa)* (Washington, DC, 1969).

——— "The US and the impending energy crisis", report, 9 March 1972 (Declassified Documents Reference System: CK3100516177, 1972).

——— "Secretary Henry Kissinger interviewed for Business Week magazine", *Department of State Bulletin* 72/1857 (1975), pp. 97–107.

——— *Foreign Relations of the United States (FRUS): 1950, Volume I (National security affairs, foreign economic policy)* (Washington, DC, 1977).

——— *Foreign Relations of the United States (FRUS): 1958–1960, Dwight D. Eisenhower, Volume XII (Near East Region, Iraq, Iran, Arabian Peninsula)* (Washington, DC, 1993).

US Energy Information Administration, *International Energy Outlook 2002*, DOE/EIA-0484 (2002). Available at www.eia.gov/forecasts/archive/ieo02/pdf/0484%282002%29.pdf.

——— "How dependent are we on foreign oil?", last updated 10 May 2013. Available at www.eia.gov/energy_in_brief/article/foreign_oil_dependence.cfm.

——— *Annual Energy Outlook 2014*, DOE/EIA-0383 (2014). Available at www.eia.gov/forecasts/aeo.

——— *Monthly Energy Review: May 2014*, DOE/EIA-0035 (2014). Available at www.eia.gov/mer.

US Federal Energy Administration, *Project Independence Report* (Washington, DC 1974).

US National Energy Policy Development Group, *National Energy Policy* (Washington, DC, 2001).

US National Security Council, "Long-range US policy toward the Near-East", draft statement of policy, NSC 5801, 10 January 1958 (Declassified Documents Reference System: CK3100451753, 1958).

Uslaner, Eric M., "Energy policy and free trade in Canada", *Energy Policy* 17/4 (1989), pp. 323–30.

Venn, Fiona, *Oil Diplomacy in the Twentieth Century* (Basingstoke, 1986).

——— *The Oil Crisis* (London, 2002).

Verbong, Geert and Derk Loorbach, *Governing the Energy Transition: Reality, Illusion or Necessity?* (New York, 2012).

Vernon, Raymond, *Sovereignty at Bay: The Multinational Spread of US Enterprises* (New York, 1971).

Vietor, Richard H.K., *Energy Policy in America since 1945: A Study of Business-Government Relations* (Cambridge, 1984).

Vitalis, Robert, *America's Kingdom: Mythmaking on the Saudi Oil Frontier* (Stanford, CA, 2007).

Vogl, Frank, "US misses its chance to influence OPEC", *Los Angeles Times*, 12 December 1978.

Waltz, Kenneth N., "The myth of national Interdependence", in Charles P. Kindleberger (ed.), *The International Corporation* (Cambridge, MA, 1970).

——— *Theory of International Politics* (New York, 1979).

Walzer, Michael, "Moral ambiguities in the Gulf crisis", *New Republic* (28 January 1991).

Weisberg, Jacob, "The corn is high", *New Republic* (12 November 1990).

Weldes, Jutta, *Constructing National Interests: The United States and the Cuban Missile Crisis* (Minneapolis, MN, 1999).

Weldes, Jutta et al., "Introduction: constructing insecurity", in Jutta Weldes et al. (eds), *Cultures of Insecurity: States, Communities, and the Production of Danger* (Minneapolis, MN, 1999).

Wendt, Alexander, "Anarchy is what states make of it: the social construction of power politics", *International Organization* 46/2 (1992), pp. 391–425.

——— *Social Theory of International Politics* (Cambridge, 1999).

The White House, *A National Security Strategy for a New Century* (1998). Available at www.au.af.mil/au/awc/awcgate/nss/nssr-1098.pdf.

——— *The National Security Strategy of the United States of America* (2002). Available at http://georgewbush-whitehouse.archives.gov/nsc/nss/2002.

Wicker, Tom, "Stating the obvious", *New York Times*, 12 January 1975.

Wilson, Caroll L., "A plan for energy independence", *Foreign Affairs* 51/4 (1973), pp. 657–75.

Wirth, Timothy E., C. Boyden Gray and John D. Podesta, "The future of energy policy", *Foreign Affairs* 82/4 (2003), pp. 132–55.

Yale Center for Environmental Law and Policy, "Yale environmental poll", Environmental Attitudes and Behavior Project, Yale Center for Environmental

Law and Policy (2007). Available at http://envirocenter.yale.edu/uploads/epoll/YaleEnvironmentalPoll2007Keyfindings.pdf.

Yergin, Daniel, "US energy policy: transition to what?", *World Today* 35/3 (1979), pp. 81–91.

——— "Thirty years of petro-politics", *Washington Post*, 17 October 2003.

——— "Ensuring energy security", *Foreign Affairs* 85/2 (2006), pp. 69–82.

——— *The Prize: The Epic Quest for Oil, Money & Power* (New York, 2008).

——— *The Quest: Energy, Security, and the Remaking of the Modern World* (New York, 2011).

Zaretsky, Natasha, *No Direction Home: The American Family and the Fear of National Decline, 1968–1980* (Chapel Hill, NC, 2007).

INDEX

www.ingramcontent.com/pod-product-compliance
Ingram Content Group UK Ltd.
Pitfield, Milton Keynes, MK11 3LW, UK
UKHW020205290926
15470UKWH00001B/21

9 781784 531